LEARNING THROUGH SIMULATION GAMES

Learning Through

SIMULATION
GAMES

BY Phillip H. Gillispie

PAULIST PRESS

New York, N.Y., Paramus, N.J., Toronto, Canada

Library of Congress
Catalog Card Number: 73-85728

ISBN 0-8091-1809-2

Designed by Dawn Hille

Published by Paulist Press
Editorial Office: 1865 Broadway, New York, New York 10023
Business Office: 400 Sette Drive, Paramus, New Jersey 07652

Printed and bound in the United States of America

TABLE OF CONTENTS

Preface

LEARNING THROUGH SIMULATION GAMES is a broad overview of a new and, I believe, exciting educational technique. In the following pages you will be introduced to relevant concepts about the technique of gaming, suggestions about how to use games, and a number of particular simulations. The concepts, suggestions, and descriptions have been penned by me and they represent my biases, my strengths, and my weaknesses.

However, the book would not have been possible without the aid and comfort of a number of persons. Larry Coppard first introduced me to gaming a number of years ago, and a group of Baptist youth from Sacramento, California, provided the context for me to originally test out this "new fangled" technique.

After I began to do more intentional work in simulation games in graduate school, my colleagues, Eric Clark, Nora Sutherland, and Steve Greenfield provided the essential ingredients of a working team relationship. They have been business associates, but more importantly, they have been trusted friends.

And finally, a great deal of the credit for the work that follows must be given to my family. Daphne, Martin, and Mark worked diligently to create the type of family environment that gives me emotional support and a reason to be concerned about the future of education.

In the deepest sense the opportunity to work on this book has been a pleasure and a privilege. I appreciate the chance to try to *communicate my love* for *peace* and *freedom,* and to be able to state that a *life* of *happiness* is dependent upon commitment to these Christian concepts.

General Introduction

About five years ago, at a youth conference, I played my first simulation game. With the enthusiastic participation of eighteen senior high students and the help of a couple of consenting adults, I administered the game, *Plans*, manufactured by the Western Behavioral Science Institute. As simulations go, this particular game is relatively complex and difficult to administer. However, the gaming experience was successful not only from the point of view of the participants, who learned a great deal about the lobbying interests in Congress, but also from my point of view as game administrator. It taught me that simulation games are not only worthwhile teaching techniques, but that they can be used by a person without a great deal of technical training.

My first reaction to a number of simulation games was "How on earth am I ever going to be able to understand all the variables—it looks so complicated." There is no way to explain away this aspect of simulation gaming. The fact that a good game model organizes and categorizes these variables into workable units is something that can be experienced only by playing a game or two.

Perhaps the best way for you to make use of this book as a resource would be for you to leaf through the book quickly and read the purposes of some of the games that are outlined. If you find a game whose purpose corresponds to something you can use, I would suggest that you obtain the game and play it with a group of high school students. The best introduction to simulation games is not reading a book, but playing the game.

What Are Simulation Games?

There are as many definitions of simulations as there are people using them. John Raser of the Western Behavioral Institute, in his popular book, *Simulation and Society*, defines educational simulations as "attempts by theorists to construct operating models of complex social and physical systems" (p. x). Raser covers the theory of simulation and simulations as theory builders or testers in depth, but only briefly suggests their educational possibilities. Sarane Boocock, a pioneer in educational simulations at The Johns Hopkins University, defines simulations as "an operating model, reproduction, or imitation of physical or social phenomena, consisting of a set of interrelated factors or variables which together function in essentially the same manner as the actual or a hypothetical system" (Sarane Boocock, "Instructional Games," in an article prepared for the *Encyclopedia of Education*, July, 1968, p. 2). She has introduced the functional aspect of simulations as imitators of social systems for learning purposes.

William Nesbitt, in an article published in "New Dimensions" to acquaint social studies teachers with simulations, reminds us of their limitations when he says, "a simulation is a selective representation of reality, containing only those elements of reality that the designer deems relevant to his purposes" (*Simulation Games for the Social Studies Classroom*, 1968, p. 4). For the purposes of this book, we will expand upon these basic definitions of simulations to include the concept of gaming.

We define simulation games as attempts to devise an environment for participants or players that they would not ordinarily experience—an environment that abstracts from reality those social, economic, or political phenomena that, together, make up a complex and sometimes confusing situation, but when reduced by simulation, become comprehensible, revealing and educational in the broadest sense.

The essence of simulation is the artificial simplification of the universe in order to single out a few variables. Yet there are few situations in real life where only a few variables are at work. Therefore, let us deal here with the complex question constantly confronting gamers, namely, how close do simulation games come to reality? The question is whether or not a game provides a realistic environment in which participants will act as if their behavior will have important consequences. Game involvement is not as intense as "real life" involvements. Nevertheless, it can be a substitute for experiences people would not otherwise have. Researchers provide several examples of the intense involvement of players to substantiate the importance of this aspect of simulations (c.f., Raser, p. 146).

Erving Goffman has suggested that the degree of involvement in a situation is not a function of the objective "reality" but rather of the "psychological reality" of the setting. He supports the feeling that a simulation designed specifically to increase the psychological reality may frequently be more involving than its real-life counterpart (*Encounters: Two Studies in the Sociology of Interaction*, pp. 27-28).

3

There are distinctions between simulations and games. Simulations are structured learning models designed to teach a specific concept or concepts about a certain system. All players, regardless of how they play, will experience the same lesson. Games, on the other hand, are not usually as tightly structured and are not as preoccupied with presenting certain concepts, but depend more on the participants' actions and reactions.

The spectrum ranges from tight simulations to loose games. However, simulations and games can be united into a single teaching technique called simulation games. The simulation provides an objective model and a framework within which participants can be involved in an exciting game mood.

In the best simulation games, this mood allows participants to examine complex systems of interaction in their complexity, rather than as isolated entities, to engage in rational decision-making procedures, and to approach problems from the perspective of several disciplines at the same time. Emphasis is placed on the development of analytical approaches and organized concepts, transferable to other problems at other times.

Thus, simulation games can provide students with the opportunity not only to become acquainted with the social, political and economic problems of our society, but with the kinds of intricate processes that go into dealing with these problems in terms of possible solutions.

Why Use a Simulation Game?

I first became excited about simulation games because the technique contains two important educational qualities.

The first quality is the ability to take the positive features of group dynamics and focus a group's energies on a particular task or a specific concept of social change. I have always been very impressed with what people learn about themselves and their relationships with others through the process of group dynamics. The human interaction and human understanding that can take place within a group context is both inspiring and educational. However, understanding one's self and the ability to perceive others within a group context as fellow humans need not be an end in itself. As many behavioral scientists who use the technique of group dynamics have indicated, individuals and groups must move beyond themselves in order to have an effect on their environment. Moving beyond one's own group into society to create positive social change is, I believe, a basic element of Christian faith.

The techniques involved in simulation gaming make use of the positive aspects of affective awareness and group dynamics, enabling individuals and groups to move beyond themselves to understand the nature of some of our social problems and to provide the tools with which to respond to these problems.

4

A second educational quality that simulation games offer is that of providing a context in which all participants can be teachers and learners simultaneously. So many structured learning situations designate one or perhaps a few people as teachers and the other participants as learners. In the gaming context, the participants teach one another and learn with one another. The role of the "teacher" is to provide the context within which students can learn from each other. Therefore, gaming is a participatory tool—it provides not only motivation but also the structure within which all game players can participate equally in the educational process.

Some people consider the motivational aspect of simulation games to be one of the greatest attributes. Although gaming does motivate participants to be highly involved, it is also important that the motivation should not be inspired totally by competition. While a few simulation games are set up on the "win-lose" model, the better games are designed to minimize this factor. Ideally, the motivation on the part of the participants should come from a cooperative desire to accomplish the specific task outlined in the purpose of the simulation game. This is not to say that competition is bad and should be totally discouraged—it's merely to say that motivation to learn a particular set of data should be based on more than the competitive stance of one player or team against another player or team.

A final word on why simulation games should be used: they're fun!

How To Use Simulation Games

The easiest and probably the best way to begin using simulation games is simply to play one of the games on the market that fits an objective of your class or group. Most of the games that are described in the subsequent chapters are neatly-packaged simulation games designed to accomplish a particular purpose, with a particular number of people, in a particular environment.

In most cases, you will be able to read the directions and administer the game without having previously played it yourself. Obviously, games vary in many ways, but a majority of the games that are introduced in this book are easily adaptable to the classroom setting.

Although there is value in playing and re-playing the same game, it is often more valuable to adapt a particular game to your own unique use. John Raser feels that the power of simulation games lies less in their actual playing than from the ability to construct and redefine them. He urges that students first be allowed to study games in general and then actually design, play, re-design, re-play, etc. This will cause students to go out and search for the information they need to perfect the game and will ultimately lead the student to believe that creating games may be a more powerful learning experience than actually playing them (*Simulation and Society*, p. 132).

Raser's point leads to the primary objective of this book: to show that it is relatively easy to design and use your own simulation games.

Some of the games that are discussed in this book were developed by my colleague, Steve Greenfield, and myself for the purpose of creating a re-usable teaching tool, that is, a packaged and marketable simulation game. However, other games were designed and used for a particular purpose at a particular time and should not be repeated. VALUES/YOUTH CULTURE was designed as a planning exercise for a weekend retreat. It is included in this book to demonstrate that one can use simulation game designs to illustrate abstract theory. It further illustrates that designing your own games can be easy and fun. (For further discussion on game design, see Appendix One in this book.)

Whether you use a "canned" simulation game, adapt someone else's game or design your own game, one rule remains constant: when used for educational purposes, simulation games should be seen as an integral part of a broader educational package. Games are very important educational tools, but they are not complete enough to stand alone. Whether the game is supported by discussion, additional reading or films, the value of a simulation game increases when used with other educational resources.

How To Use This Book

LEARNING THROUGH SIMULATION GAMES is designed as an educational Sourcebook for use in Junior, Senior High, Colleges, Universities and Adult Education settings. The sourcebook has been conceived as an instrument for making available to the teacher a wide variety of data which can be used in a variety of subject areas.

Although suggestions will be made as to whether a game is more appropriate for certain settings, the final decision should rest with the educator. After reading a brief description of the game, analyzing its objectives and the readiness of the class, the educator is the best judge of which simulation games to use.

The games that are discussed in this book cover a very wide variety. In the following pages, you can read about a computer simulation that takes a minimum of a full day to play on an I.B.M. 1130 computer. You can also read about an hour-and-a-half parlor game; about a game based on as abstract a concept as value systems. You will also find a game based on the concrete problem of how to choose the right "life career." The games described range from the consideration of complex social problems, such as racism and peace, to the equally complex personal problems, such as, "How do I communicate with my parents?"

Despite the variety, each game is discussed through the use of a common outline. That outline is The Name of the Game: General Information; Purpose; Components; The Players; Environment; A Scenario. Under **General Information** you will find the cost, playing time, number of participants, and where you can order the game. **The Purpose** gives a brief explanation of the objectives the game hopes to achieve. Under **Components,** I will describe the various systems that make up the game, e.g., in the game, Ghetto, the systems that are built into the simulation are employment, education, hustling, and welfare. **The Players** will tell you who, how many and how sophisticated. This section will also say something about the type of player activity that is expected. The section on **Environment** will set forth the physical as well as the psychological setting of the game. **A Scenario** is not included in each game, but is usually a part of the first game described in each chapter.

This book can be a resource not only for the games described in the following pages, but also for other games that you will come into contact with. After playing a few games, and reading about a number of others, you will be able to make evaluations about the games you see and read about. Does it have enough action? Are the time periods accurate? Can it be administered by one person? These are questions that you will be able to answer for yourself about "new" games when you have played eight or ten games from this book, read about a dozen more, and tried to write a game or two of your own.

Theme One: Freedom

But what is freedom rightly understood
a universal license to be good.
—From *Liberty* by Hartley Coleridge

The concept of freedom is simple. The desire to be free is intense but the reality of freedom is elusive.

Too often people view freedom only in an individual or personal sense. Freedom is seen as "my being free." Obviously the opportunity to experience individual freedom of choice is essential in our society. It is also obvious that we need to *feel* free from oppression. But what is not so obvious is how to achieve freedom.

The thesis of this chapter is that one means to the achievement of freedom in our society is the understanding that freedom is more than individual freedom—it means more than "feeling free." In order to construct a free society, men and women must understand and then change the institutions of our society (that is, the economic, political, social and educational systems) that separate men and women into different classes and different groups.

THE NAME OF THE GAME:

Urban Dynamics

Order from _____ Urbandyne
5659 S. Woodlawn
Chicago, Ill. 60637

Cost _____ $95.00

Playing Time_____ 4 hours +

Number of Participants _____ 12-20

PURPOSE:

Urban Dynamics is designed to give students an opportunity to build a city. The players begin with a stereo-typical urban area of 1920 and build the city with residential units, economic units, and employment units. They establish the political and educational systems for the city and then make the decisions that keep these systems running. The participants are the homeowners, the taxpayers, the politicians, the corporation presidents, the factory workers and the poverty stricken. The game *Urban Dynamics* gives the players the chance to experience the dynamics of some of the major social problems.

The game can be played in two versions: the historic version and the future version. In the historic version, the participants learn the way in which social problems developed. For example, they learn how center city areas deteriorate because of the declining economic conditions; why the great rush to the suburbs; and why the tax base for most urban areas is being eroded. In other words, they obtain some historic perspective on the major social issues that are facing our urban areas today.

In the future version, the game begins in the 1960s. Ideally the future version should be played after the participants have been through the historic version of the game. Beginning with the 1960s, the players are asked to create solutions to some of the social problems that have developed in our urban areas; using the same economic, political, educational and social tools, the participants are challenged to overcome the problems of poverty, powerlessness and economic deterioration.

COMPONENTS:

The game players have the chance to create their *social* environment. The playing board represents the land area of the city. Various parts of the playing board are blocked off or owned by a particular

11

social class, for instance, higher, middle or lower. One-inch square blocks are used to denote population units, which are placed on the playing board by the game players and eventually become neighborhoods where the participants live. The type of residential area that is created is dependent upon the decisions made by the game players.

The *economic* sector of the city is also developed by the players. Depending upon the resources of the various teams, factories or corporations may be purchased and run at a profit. These factories and corporations which are owned and controlled by the participants also employ residents of the neighborhood. In the game there are two levels of employment: white-collar and blue-collar. Within the economic sector, the game provides tensions between management and workers, between the employed and the unemployed, between business interests and political interests.

The most exciting aspect of the game is the opportunity participants have to run the *political* sector of the city. The neighborhoods are divided into councilmanic wards, all of which are represented on city council. The participants form a city council which elects the mayor, sets the tax rate, and makes decisions concerning welfare, education, urban renewal and urban redevelopment. The interrelationship between the political, economic and social sectors becomes quite apparent as the city council attempts to set the tax rate and decide the level of welfare to the unemployed.

The *educational* sector is the weakest of the four major components of the game. The only real contribution that education plays in this game is the requirement that an employment unit must be educated in order to obtain a white-collar job. Education becomes a political tool in the game because the city council must appropriate money for education and thereby decide how many employment units will become available for white-collar jobs.

THE PLAYERS:

The game players consist of the game director and the participants, divided equally into four teams. The four teams represent four social classes. All the game parts are color-coded by team; for example, the yellow team places yellow population units, buys yellow economic units, and employs yellow employment units. It places its units on land that is owned by its team.

The yellow team represents the landed aristocracy. It owns the land along the shore—the gold coast. It begins the game with a number of economic units and a large amount of liquid cash. The white team represents primarily middle-class America—this includes the working class as well as most elements of the "professional" class. It owns some land and has some economic assets, but its primary source of power is its dominance in the political sector of the city. The black middle class is represented by the blue team, which has some money and some economic power, but both are very limited. The political clout of the blue team is practically non-existent. The red team, which represents the poor people of the city, has nothing—except population.

12

ENVIRONMENT:

Everything needed for the playing of *Urban Dynamics* is contained in the attaché case which can be purchased or rented from *Urbandyne.* The game does not require additional space or other special effects.

The game board should be set in a semi-circle around the board. The director of the game, situated in a central position, will begin instructing the students concerning each move they make. As the game progresses, the director will find that he or she will be moving away physically, as well as psychologically, from a central role in the simulation.

The introduction to the game should take about thirty minutes. This includes introducing the students to their roles; explaining the purpose of the game; and explaining the initial game set up for the 1920s. The leader's manual, provided with the game, will explain how the game director leads the participants through the 1930s, explaining how, when and why each team makes each move. The leader should allow 15 to 20 minutes for the 1930s "walk through."

The participants take over the game and begin making decisions with the 1940s. The leader should allow one hour per round/decade. A strict time limit should be set because one of the factors in the simulation is operating within the time limit.

If possible, it is better to play the game in one four-hour sequence. However, it is designed in such a way that it could be used in one-hour segments. For example, the introduction and "walk through" for the 1930s could be completed in one class period. Further class periods could be devoted to each successive round. The historic version of the game usually runs through the 1960s. Actual playing time, therefore, should be considered to be approximately four hours.

A SCENARIO:

The game, *Urban Dynamics*, has been used successfully by groups from senior high age through graduate students. The key to making effective use of this game as a teaching tool is to understand clearly the objectives of the game. These objectives are: 1. To illustrate how, under our present economic, political, social and educational systems, the gap between the rich and the poor has continued to widen. 2. To demonstrate the power and the influence of the economic system over the political and social systems, especially with regard to attempts to distribute power and freedom to the disenfranchised. 3. To indicate that much of the denial of freedom in our society is unintentional and due to outmoded, ineffective and often unfair political, economic and social systems that need to be changed and updated to meet the needs of modern society.

I believe the best way for the reader to gain a better understanding of how the game serves to achieve these objectives is to discuss the experience my colleagues and I have had with this game over the past two years.

The most obvious thing that happens in every run of the game is that the blue and red teams continue to expand around the core of the city, and the white and yellow teams begin to flee to the suburbs in greater numbers. Also, as the city tax rate rises, the white and yellow teams begin to build their factories and corporations in the suburbs to escape the high tax rate of the city. Therefore, as the game develops, so does this "city"—in almost exactly the same way every major city in this country has over the past thirty years.

The yellow team, which begins the game with economic superiority, maintains that superiority throughout the game. The psychological effect of wealth on the participants of the yellow team has been particularly significant. Although a number of people who have played on the yellow team are known in their private lives as liberal advocates of social change, without exception, they devised ways of giving money or land to the blue and red teams in a manner that allowed them to maintain control over those two teams. It is probably incorrect to say that, at any time, the yellow team "gave" anything to the blue or red team; it would be more correct to say that, at strategic times, they made concessions based on their own self-interest. Even though this is a game, the team's own self-interest was the highest priority. The yellow team members also felt that the lot of the blue and red teams was improving and that they were responsible for the improvement. They cited, for example, the new factories and corporations that they built in order to employ people of the blue and red teams, forgetting to point out that their profit from these corporations far exceeded the amount that they were paying to the employees. At no time have groups thought of profit sharing, or even the most elementary type of distribution of wealth. Always at the end of the game, the gap between the yellow team and the red team is far greater than it was at the beginning of the game. So, even though the red team had made some improvement, the yellow team had improved to a far greater extent.

Although economic power is always in the hands of the yellow team, political power usually rests in the hands of the white team. Occasionally by the 1950s, the red and blue teams together could be the balance of power between the white and yellow teams, but the significant factor is that it doesn't make much difference. Although political decisions are emphasized a great deal in this game, it is really the economic decisions that alter the course of the game. Real power comes not in controlling the city council, but in being able to buy new corporations and new factories. Different coalitions developed each time we have played the game, but at no time were the white, red and blue teams able to coalesce to challenge the economic power of the yellow team. Also, the yellow team was able to maintain its economic control more effectively when it did not spend a lot of time and effort trying to control the political aspect of the game. In other words, when the yellow team spends a great deal of time and effort working out a coalition with the red and blue teams, in order to maintain control of city council, the economic advantages lost far outweigh the political advantages gained.

Only one time did such a coalition take place. The yellow team, in conjunction with the blue and red team, gained control of city council. In return for political support, the yellow team bought property for the red and blue teams to set up their own suburban community. The yellow team considered this a magnanimous effort but forgot to realize the red and blue team had no money for industrial development in their suburban community and wound up economically worse off than when they began. The primary factor, therefore, in this game is the idea that a favorable economic climate is the main factor affecting the political decisions. In other words, if the tax rate goes too high, corporations and factories will move outside the city. And since it is the industrial sector that contributes most to the city treasury, it is imperative that the economic climate remain favorable to the yellow team. However, the extenuating circumstances, such as expanding population of the red and blue teams, and limited land area in which to develop, make it more and more difficult for the city council to maintain the favorable economic climate. Therein lies the dilemma of the game, and therein lies the dilemma of every major urban center in the United States.

Moving from the economic dilemma of the city to the psychological effects of this game upon some of the participants, it has been interesting to note that the members of the red team very quickly lose interest in the game. On one or two occasions, they almost completely dropped out—they had to be reminded when it was their turn to make a move. By contrast, the members of the white and yellow team work at a very high level of involvement and are seldom caught sitting by talking idly about something unrelated to the game. In fact, it is usually difficult to terminate the game because of the protests of the yellow and white teams; they want to go on with the game because they have great plans for the next round.

The primary strength of this particular simulation is that it gives a student or participant an overview of the way a city develops. It demonstrates visibly the way various human, economic, political and social components have interrelated to bring about some of the problems we have today. But, in addition to teaching the overall concepts of city development, the game also has the potential for profound impact upon individual participants. For example, when we played this game with faculty members from a state college, one of the participants was the head of the campus security force. Two weeks later, when we were doing an overall evaluation, the head of the security force said that the *Urban Dynamics* game had had the most profound effect upon him. He went on to say that as he was driving home later that evening a thought suddenly hit him like a bolt out of the blue. He said, "Here I am, a professional policeman, a man trained in police work for eleven years, a man conservative by nature, suddenly thinking to myself that the only alternative for the blue and red teams is open rebellion." He and others went on to elaborate that the system is so tight with the economic and political systems as they are presently devised, there is no way that those of the lower economic classes can gain power (i.e., become free).

While the game points out vividly the economic class structure that is present in this society, it does not guarantee that the participants will question the values of this class structure. In other words, as we continually raise the question about the value of one team or one class holding and maintaining economic power, the participants often resorted to the argument, "That's the way it is" and "That's the system we have to work with." It was quite disturbing that more people did not question the underlying value structure which supports the type of economic class system that contributes so substantially to our urban problems. However, there is some indication that the immediate reactions to the game may be quite different from the long-term reactions. For example, one college professor who played the game as a member of the yellow team, in the evaluation immediately following the game, defended his role as a benevolent benefactor of the red and blue teams. He also pointed out at great length that the economic system is flexible and that, with a little help from the yellow team, the red and blue teams would eventually be able to make it. Another professor who played the game mentioned two days later that he had since talked with the first professor and discovered that the professor who was so defensive about the status quo had stayed up until 4 A.M. in the morning, reflecting on the role that he had played, and the way he had played it. He indicated that he was seriously re-thinking the way he had played the game, and was questioning his immediate reaction to the game. The same professor also indicated to me three days later that, although he had at first thought the game had little value, he now felt it had immense value as a teaching technique. He was also convinced of the need for economic social change in the urban areas.

The evaluations of this game reinforce our feelings that the objective evaluation of the game should take place a day, maybe even a week, after the actual playing of the game. This gives the participants more time to reflect on their opinions, more time to reflect upon their ideas about the way the game developed. The involvement of the participants has usually been at a significantly high level to assure that there will be a good deal of post-game reflection by them.

In addition to its value as an educational tool, the game has a definite mechanical advantage. It can be carried in a briefcase and operated by one person. Add this advantage to the fact that *Urban Dynamics* is a good, tight simulation and is an exceptionally good teaching tool.

NOTES

THE NAME OF THE GAME:

Clug

Order from	*Complete Kit* Urbex Affiliates, Inc. 474 Thurston Road Rochester, New York 14619 *Player's Manual Only* Tu Free Press Dept. F Riverside, New Jersey 08075
Cost	$75.00 (Complete Kit) $4.95 (Player's Manual)
Playing Time	10-15 Hours
Number of Participants	3-30

There are many variations of the original model of *CLUG* which can be obtained from Cornell University. I recommend using the variation described in *Simulations on Systemic Racism: Variation on the Community Land Use Game*, by Robert A. Spencer, which is distributed by the Interreligious Center for Urban Affairs, Shell Building, Suite 811, 1221 Locust Street, St. Louis, Missouri, 63103.

PURPOSE:

Professor Allan Feldt, of Cornell University, writes in the introduction to the *CLUG* Manual:

> "The Community Land Use Game (CLUG) attempts to reduce the broad range of variables supposedly affecting urban land-use decisions to a small number of basic attributes of cities and their surrounding territories; by making decisions about how these basic components are to be employed, players build, operate and maintain their own community."

This brief and cogent purpose describes what, in my opinion, is the single most important contribution to social-science simulation games.

CLUG was developed in 1965 and provides the model for a number of simulation games now in wide use. But just as important as the adaptations of *CLUG* are the numerous variations of the original game.

Each adaptation and variation provides its own specific thrust, with its own specific purposes. The original *CLUG* was designed to teach the basic economics of land use.

COMPONENTS:

CLUG is played on a board consisting of a 14″ x 14″ matrix of squares. Major highways, terminals, and utility plants can be placed along any lines on the board, and any line that is not otherwise designated is assumed to be a secondary road.

Industries, stores, offices and residences are the primary types of land use recognized in *CLUG*. There are two types of industrial land— full of industry and partial industry. The designation of partial industry refers to an industry equal to one-half of a full industry. Residential units may be established in any one of four different densities. The residential construction cost per unit increases with the degree of intensity of residential usage. There are two types of stores used in *CLUG*. The offices are used to designate executive work that is done for the industries and the local and central stores. They represent such diverse activities as bookkeeping and legal representation.

In order to develop any square of land in the game, utilities must be provided by the city council. Utilities represent a broad range of municipal services, such as water and sewerage, electricity, police and fire protection, and education.

The decisions concerning taxation and the extension of utilities are made in the political sector of the game. The participants form a city council and set policy concerning taxation and provision for city services.

ENVIRONMENT:

CLUG is the type simulation game that one can play again and again. It has the excitement of a Monopoly game and the educational value of an excellent economics text.

The game is not difficult to administer. In fact some would argue that it is easier to administer the game than to play it. A minimum of two people are required to administer *CLUG:* one person to direct the step-by-step play of each round and another person to do the computation.

After you become familiar with the basic *CLUG* model, I strongly advise using a variation which weights the resources of the various teams. As I mentioned above, the best description of this type of variation is available from the Interreligious Center for Urban Affairs, in St. Louis. The weighting of the teams produces a dynamic very similar to that present in *Urban Dynamics*. An economic class system is usually very quickly reinforced and perpetuated in this variation of *CLUG*.

When one begins to explore the numerous variations on *CLUG*, the prospects are practically limitless. There are now at least three computer-assisted versions of *CLUG* available. The most usable computer-assisted version that I am aware of was developed by Complex, an industrial mission in Raleigh, North Carolina.

In addition to developing a computer-assisted variation that can be operated from a time sharing terminal, Complex has also made creative use of video tape recordings with *CLUG*. When the television media is added to the game, a separate team is designated to "report the news of the developing city." Each round, the media team circulates through the other teams, gathering data and news items. At the conclusion of the round, the media team presents a three-to-five-minute "newscast" on the events in the developing city.

CLUG, like *Urban Dynamics*, should be played in long blocks of time. It is ideally constructed for a weekend retreat; however, it is possible to play a round a week for six to eight weeks. When the game is used over an extended period of time, it is suggested that the participants do a continuing evaluation.

CLUG is exciting; *CLUG* is educational. *CLUG* is a must for persons deeply involved in social-science simulation games.

NOTES

THE NAME OF THE GAME:

Metropolis

Order from ———————— Urbex Affiliates, Inc.
474 Thurston Road
Rochester, N.Y. 14619

Cost ———————— $200
(Players will have to use
IBM1130 or 360 computer
and computer operator)

Playing Time———————— 6 hours

Number of Participants ——— 7-45

PURPOSE:

The most complex of the social-science simulation games is the computer-based game. *Metropolis*, developed by Professor Richard Duke of the University of Michigan, was the first such game. It is now nearly seven years old, and in my opinion is still number one in its field. *City I* and *Apex* are more recent games built on this model and both originate from Duke's Environmental Simulation Laboratory at the University of Michigan.

Metropolis is run on an IBM 1130 or 360 computer and the program can accommodate five groups working simultaneously. Each group has three basic roles: politicians, real estate speculators, and city planners. (These will be described in greater detail below.) Each of the five groups works with the same variables—each team is responsible for running a city.

The players decide on building programs, streets, schools, taxes, zoning—they hold elections and evaluate politicians, and land speculators. Each team has its own "scoring" but all must work together in some fashion for all to score well and for the city to be successful . . . The city has a detailed background on population growth and make-up, geography, socio-economic pattern, and so forth. This information is available to the players from tables, charts, descriptive memos, and a detailed map of the city.

Decisions are made once a "year" in the game-city time. Then a computer takes over and determines the impact over the year of these decisions and other typical events. It notes changes in the city's population, its economic status, its social condition, and the life. The computer produces these details for each team to review as it works toward the next year's decisions. Even a city newspaper is computer-produced to assist the teams in this highly realistic process of running a city ("Introduction to Metropolis," an unpublished paper by *Techtonics*, Ithaca, New York, 1963, p. 12).

COMPONENTS:

The city that is simulated in this game is East Lansing, Michigan. It is a city of about 200,000 and is divided into three wards. There are three *politicians* in each of the five city-groups. The politician from the first ward represents a largely black, poor neighborhood. This ward houses the government buildings and much deteriorated housing. It is overcrowded and there is much agitation from civil rights groups.

The second ward by contrast is composed largely of white, working class families. They are in a higher income bracket and have better homes than the residents of the first ward, but their economic plight leaves much to be desired. They are, for the most part, hostile to the demands of the residents of the first ward, but are actively seeking more improvements for their ward.

In the third ward, the residents are mostly middle class and upper middle class. A university is located in this section of town, as well as some of the city's best recreational facilities. On the conservative to liberal spectrum, the residents of the third ward are more liberal than those of the second ward. They supported progressive legislation in city council, such as welfare reform and new schools for the first ward.

The politician's team in each group appoints one of its members councilman from the first ward, another member councilman from the second ward, and another councilman from the third ward. The politicians must make a single set of decisions each round, but they are rated individually on their ability to please the residents of their respective wards. Every two rounds, an election is held to see if the politicians are re-elected.

A second team is the *real estate speculators*, who invest by buying land in one or more of the three wards. They may purchase the land for residential, commercial or industrial use. The computer simulates the residential growth of the city, and each round the speculators' printout states where growth has taken place and how much their investments have grossed. Based on information from the planners and politicians as to where new city construction is going to take place, the speculators buy new land. Their ability to predict where there will be new growth and construction determines whether or not they will increase the value of their investments.

The third team is the *city planner/administrators*. They are given the task of planning the future needs of the city. Their decisions are made one round ahead because they are made in the form of recommendations to the politicians. From a list of 106 various capital projects, the planners recommend in round two the improvements they feel must take place in round four. They must keep the residents of the city happy, but also get some innovative projects through. They are evaluated on the basis of their ability to convince the politicians and the community-at-large (simulated through a public opinion poll) of the necessity of these improvements.

In addition to the decisions of the other teams, each team is affected by certain national and state economic trends that are built into the computer program. For instance, the city is largely dependent upon the automotive industry for its economic growth. During the third or fourth round, a national recession takes place and there is a sharp drop in automobile sales. This affects the entire economy of the city and all teams must make adjustments by trimming their spending. The politicians often find themselves deeply over-committed in terms of new city capital improvements and, therefore, must do nothing for a year or two. This, of course, hurts their chances for re-election. The speculators have a lot of money tied up in land that is not being developed and, therefore, cannot use it for future developments. The clues to the economic crisis are carried in the newspaper that comes out each round. It carries the only indicators of outside influences on the city, but it also carries news pertinent to the decisions being made about the city's internal affairs.

Another factor that affects each of the teams is the *public opinion poll* that is administered each round. "The issues usually suggest some sort of project and expenditure activity—they also reflect some of the outside economic influences on the city" (Ibid., p. 7). All teams are affected by the results of the Public Opinion Poll simply because it represents public opinion.

ENVIRONMENT:

Each round is begun with each team making decisions on the public issues (Public Opinion Poll). After this, each team has a longer list of decisions to make.

> Administrators make decisions as to what projects are to be recommended for the following year—in effect, the second year of a five-year capital improvement plan. Politicians decide what to spend this year—what projects are actually to be undertaken. Speculators indicate where and how much they want to invest (Ibid.).

Each team's decisions must take into account what the other teams are doing. For instance, the real estate speculators need to know what projects the planners are recommending and the politicians are authorizing in order to know where to invest. The politicians are the most independent of the three teams, but they, nevertheless, must have rapport with the other teams or they will be hurt on the Public Opinion Poll. The administrators are perhaps the most vulnerable to outside influence. They need to collaborate closely with the politicians to see that their recommendations are put into effect. Their clout comes through a coalition with the speculators. The administrators can promise to recommend projects in areas where the speculators have invested in return for pressure by the speculators on the politicians. The speculators may make campaign contributions to the politicians each round. A campaign contribution, obviously, increases the politicians' chance of being re-elected. In short, the game is one that requires a great deal of team interaction. Occasionally, even teams from different cities (groups) come together to discuss strategy.

The computer program is set for ten rounds or years, but most groups have exhausted the game or themselves by the seventh or eighth round. Playing time is five to six hours and the "talk-down" takes about an hour. One other note on the mechanics of the game is appropriate. If a group is playing for the day, it is best to serve a take-out lunch or dinner to the players and let them eat as they play. The meal becomes a realistic part of the simulation in that it provides necessary detraction from their routine, but does not provide an escape from the reality of their dilemma. If a group breaks for an hour or so to eat, it often takes another hour to get back into the game.

GENERAL COMMENTS:

Although *Metropolis* was designed primarily for use by graduate students in city planning, we have found it to be an effective tool for use with senior high students. Initially, the game looks extremely complex and unfathomable. However, once the participant has "walked through" the first round and learned to read the computer print-out, *Metropolis* is no more complicated than many of the less-sophisticated manual games.

The primary point to stress in discussing *Metropolis* is the central focus of a city's capital budget. This extremely important part of "how a city operates" is often overlooked by persons who are involved in urban social change. The relationship between real estate interests and political interests is usually quite close.

Groups that use this game are encouraged to visit and study the local government of their town or city, both before and after playing the game.

Although this simulation is of a metropolitan area, the components of the game do not differ significantly from the major components of the city government of a town or a large city.

NOTES

THE NAME OF THE GAME:

Star Power

Order from	Western Behavioral Science Institute SIMILE II P.O. Box 1023 - 1150 Silverado La Jolla, California 92037
Cost	$3.00 for instructions on making your own set; $25.00 for an 18-35 participant set
Playing Time	Approximately 2 hours
Number of Participants	18-35

PURPOSE:

The primary purpose of this simple, but sophisticated, simulation is to illustrate the relationship of power to powerlessness. The participants are divided into three teams—the squares, circles and triangles. In this three-tiered society, the teams are distinguished by their amount of wealth, proof of which is represented by chips of varying denominations. The object of the game is for the participants to improve their relative wealth through a process of bargaining and exchange.

After a couple of rounds, the members of the wealthy team are given the opportunity to make new rules for the game. The manner in which they use power and the manner in which the powerless respond to their use of power, is more than interesting. It is informative about human nature as well as society as a whole.

COMPONENTS:

The components are the resources that the individuals have as members of one of the three teams. The players who are on the "square" team are the most wealthy. They are given chips of greater value than the players on the other teams and, of course, attempt to maintain their status as members of the wealthy class.

The circles represent the middle class. They should not be content in maintaining the status quo, but should be attempting to gain enough wealth, through exchange, to become squares. The triangles are the least wealthy of the three teams, and are struggling to move up the ladder of social mobility. Each player is tagged by wearing the symbol of his team around his neck and must learn to relate to that identity and the identity of the other participants in the game.

GENERAL COMMENTS:

I doubt that any other simulation has had the widespread use and acceptance that *Star Power* has enjoyed in recent years. I usually think of it as one of the primary bridges between group dynamics and simulation games. *Star Power* has the excitement and power of individual interreaction without the psychoanalytic aspects of some of the more sophisticated forms of group dynamics. It also has the structure of a simulation game in that it attempts to set forth the systemic nature of one of our most serious social problems, namely the uneven distribution of wealth.

Star Power could be criticized for not being a tight simulation game. That is, it does not identify and then replicate a particular economic condition. However, I believe it should be used as a simulation game primarily because it does serve as an excellent introduction to the concept. Persons playing it, however, should be careful not to define simulation games using *Star Power* as a prototype.

When using this game, there are two different levels on which it can be analyzed. First, the players should examine their own response to wealth/power or lack of wealth/power. The individual participant should examine his own motives toward those who are different from himself. Therefore, the game must first be analyzed at the individual level.

But it also should be analyzed at the societal level—are there social classes in our society? Are there groups of people in our society without wealth and/or power who have the same problems achieving upward mobility as the triangles and circles had in this game? Understanding the systems nature of many of our social and economic problems is quite helpful in preparing to use some of the more sophisticated simulation games like *Urban Dynamics*, *CLUG* or *Metropolis*.

NOTES

THE NAME OF THE GAME:

Sitte

Order from ———————— Western Behavioral Science Institute
SIMILE II
P.O. Box 1023 - 1150 Silverado
La Jolla, California 92037

Cost ———————————— $3.00 for a sample
$35.00 for a 25 student kit
$50.00 for a 35 student kit

Playing Time———————— 2½ hours

Number of Participants ——— 18-30

PURPOSE:

The game *Sitte* illustrates how five special interest groups—business, the disenfranchised, government, ad hoc committee for parks and trees, and taxpayers association—interreact in their attempt to improve the condition of *Sitte*. The decisions which result in the interaction of the five interest groups or teams are rated on the following indicators: equality of reward in *Sitte*, overall aesthetic writing of *Sitte*, profits after taxes and total taxes. The participants will gain a sense of the difficulty in improving the conditions of *Sitte* because the special interests of each group often conflict with the general public interest.

COMPONENTS:

The primary components of this game are the resources available to the five interest groups listed above. The participants are divided into five teams, each team representing one of the interest groups. The *business team* is concerned with improving the "outsiders' image" of *Sitte*. It also wants to improve city services and the cultural and educational offerings, but its primary focus is to try to get the government to carry the main burden for improvements without affecting its position in *Sitte*'s power elite.

The *disenfranchised*, on the other hand, are attempting to become a part of the power structure. Their goal is to secure equality for all the citizens of *Sitte* as well as to improve city services and the cultural and educational offerings of the city. The disenfranchised represent the minority groups—poor people, young people, elderly people—all citizens who do not participate in the decision-making process of the city.

33

The *taxpayers association*'s primary concern is, of course, the rising tax rate. It wants to improve the services of the city, improve its image, but do this without raising taxes.

Sitte's *government* represents the mayor, city council, and all of the governmental agencies in the city. It is concerned with the image of the city but it also wants to improve services and obtain the help of business and other groups to carry the burden of making changes and improvements in the city.

The aesthetic appearance of the city is the primary focus of the *ad hoc committee for parks and trees*. It emphasizes educational and cultural opportunities for citizens of *Sitte* and is also concerned with questions of ecology.

GENERAL COMMENTS:

Sitte is very usable and easily adaptable for the classroom. It progresses by rounds which vary in length from 25 to 45 minutes. A teacher or simulation director could use this game in four daily sequences or it could be played in one three- to four-hour block of time.

Another distinct advantage of *Sitte* is that the participants can very easily rewrite the contents of the game to make them more applicable to their own situation. Not only is there an opportunity for the teams to write new proposals, but the teams themselves could even be redefined or renamed. I believe the game is used most effectively when a group plays two or three rounds of the game as it is presently packaged. After that, the participants should be asked to write their own game, providing content that is relevant to the city in which they live. One very useful variation on this game is adding a sixth team representing the churches or religious institutions. This team, in contrast to the other five teams, would be given no influence points. However, it does have free access to all other teams and may use "moral persuasion" to try to influence the decisions of the other teams. If you have access to video equipment, you may also add a media team that works throughout the game, gathering information and producing news shows for *Sitte*. (For a more detailed discussion of the use of video equipment, see the introduction.)

Sitte also provides for a lot of individual and team creativity. For example, it is not unlikely for the disenfranchised team members to organize a demonstration. To do this they would actually make placards and start a picket line around the government team, loudly proclaiming their demands. The government team, in response, can order that the leaders of the disenfranchised team be expelled from the game for a round or put in "jail." An endless variety of creative responses will, no doubt, be forthcoming from the students who play this game.

In short, the power of this game is its flexibility, and participants should be encouraged to take advantage of that strength.

NOTES

THE NAME OF THE GAME:

Ghetto

Order from ———————— Western Publishing Co.
850 Third Ave.
New York, New York 10022

Cost ———————— $20.00 a game or $16.00 with
an educational discount

Playing Time———————— Approximately 2½ hours

Number of Participants ———— 10

PURPOSE:

Dove Toll, the primary designer of the game says:
Millions of people in the United States live in poverty.
We all know this in the abstract. But how does it *feel* to be
poor. Few middle-class Americans are in a position to appreciate the actual life conditions of the poor. Their basic needs
are provided for. They automatically and routinely make career and educational decisions which give access to the
goods of life; and the general culture approves their aims,
motivations and attainments. Many prosperous Americans
suffer from other troubles—spiritual and psychological in nature—but they are essentially free of the economic agonies of
the ghetto.
The game of *Ghetto* was designed to give more privi-
leged Americans a small taste of some of the pressures that
work on the poor of an inner-city neighborhood.
. . . *Ghetto* can sensitize its players to the emotional,
physical and social world the poor inhabit (*Ghetto*, Instructor's Manual, New York: Western Publishing Co., Inc., 1969,
p. 3).

The designers of the game intend it for use with white, middle-
class Americans, and its primary purpose is to give the players a vicari-
ous experience of some of the pressures and agonies that are the everyday experience of ghetto dwellers.

COMPONENTS:

The game progresses in rounds. In each round all the players
must invest their total amount of hour chips (each hour chip represents

37

the number of hours per day that an individual has to improve his or her situation in life). Each investment is rewarded with a certain number of points. For example, the immediate reward for education is very small— 10 points for eight hour chips invested in high school. The immediate reward is quite high for hustling—15 points for each chip invested. In work, the jobs range from unskilled through semi-skilled, skilled I, skilled II, semi-professional and professional. The reward for each of these jobs ranges from 90 points for eight hour chips invested in an unskilled job to 300 points for eight hour chips invested in a professional job. The individual player may only invest in a job for which he has the prescribed educational requirements.

The game provides for unusual circumstances (represented in chance cards), such as illness on the job or receipt of a special honor in school. Also, during each round the hustlers have a percentage chance of being put in jail on the basis of a roll of the dice.

The game continues for a maximum of ten rounds. The object is for each of the participants to devise a means of surviving and, hopefully, getting ahead in life. However, in addition to the element of individual achievement, the game also functions at a second level—that of neighborhood improvement. After the second round, the neighborhood condition chart becomes an integral part of the game. In order to have the opportunity to improve the neighborhood, the game participants must invest their hour chips in one of the four areas (housing, education, recreation, or safety) in multiples of five. The reason the chips must be invested in multiples of five is to demonstrate that the members of the neighborhood must cooperate in trying to improve the neighborhood.

THE PLAYERS:

In the game each of the ten players represents a ghetto person with particular extenuating circumstances. For example, one person would play the role of Billy, age 16, no dependents, has completed a ninth grade education, and is single. His source of income would be specified as mother and odd jobs and he would begin the game with 12 yellow hour chips. Another person would play the role of Betty Jo, age 30, with six children. Betty Jo has completed the eighth grade, is separated from her husband, receives her income from welfare, and begins playing with four blue chips. The other eight players are given a similar profile folder of a particular stereotyped ghetto dweller, ranging from Bill, who is young without family responsibilities, to the welfare mother, who is older and has many family responsibilities.

ENVIRONMENT:

Ghetto requires no advance set-up or particular arrangements. The ten participants gather around the two-foot-square playing board, which is usually placed on a table. The game moves better and faster if the participants do not use chairs. They should be encouraged to move around the playing area, interacting with other participants.

Although the game is designed for ten persons, as few as seven and as many as twenty can play. If there are more than seventeen participants, I would suggest using two games. It is possible to double up on the roles, but that cuts down on the learning that takes place. Also, it is possible for one person to run as many as three games simultaneously in the same room.

At the conclusion of the game, the participants tally their scores in two ways. First, they total the number of points for each round and, second, they determine the highest job that they qualify for by virtue of their education, and multiply the number of reward points that they get for this job by forty. The second score is based on the assumption that an individual has approximately forty years of productivity in a particular occupation and, depending on the amount of education he or she receives during the time span of this game, certain long-term rewards can be expected. The ten players are then paired off in terms of the number of hour chips and conditions in which they began the game. Therefore, the only scores of the players that would be compared are those of the two participants who began the game in relatively the same economic and social position. The designers of the game hasten to caution against making very much of the particular scoring because so much of what happens in the game is a result of chance. However, it is possible for the two participants who began at approximately the same level to have a significant discussion about the strategy that they used in order to survive and get ahead in life.

NOTES

THE NAME OF THE GAME:

Blacks and Whites

Order from _____ Psychology Today Games
P.O. Box 4762
Clinton, Iowa 52732

Cost _____ $5.95 plus 50¢ for handling

Playing Time_____ 2 hours or less

Number of Participants _____ 3-7

PURPOSE:

When you are systematically denied whatever chips they're playing with, in whatever game they're playing, you have to adopt a strategy of semi-participation, of cooling it. You're poor and pure; you use what little clout you've got to keep the land hungry majority type from winning the game cheaply and quickly. But when things start to change and you grab some odd breaks, it's time to stay loose and invent wild new strategies. You may use resources better, and risk yourself with more courage, than players who start well-off but live up-tight in fear of failure.

The U.S. Organization of militant black men and women, consultants on developing the game, recommend that you revise the rules . . . each time you play—until you build an entirely new game (*Psychology Today* in its game instructions).

Understanding the rules so you can revise them, revising them, building a new game—when a person has done that, when he/she even begins to think like that, he/she is close to being free as an individual. The point is, of course, that when the game is the entire societal structure, it's not so easy to change the rules.

COMPONENTS:

The primary component of this game is the discrimination of whites against blacks. Through the playing of the game the participants should come to a better understanding of the racism in our society and begin to get a handle on how they personally can work to change the pattern of discrimination that exists in this country.

THE PLAYERS:

The game is designed for three to seven participants. Each individual plays the role of a black person or a white person. Some interesting dynamics can occur when more than one game is being played in the same room.

When the players exercise creativity in their responses, the game becomes more meaningful and exciting. There is not a need for one person to serve as game operator, although someone who has played before can sometimes encourage the participants to liven up the game.

ENVIRONMENT:

Employing real-estate idiom, the game uses a playing board, dice, and "opportunity cards" in an attempt to give its players a taste of the helplessness that comes from living against implacable odds.

When the game was first designed, its instructions relate, "players who chose to be black could not win, or seriously affect the course to the competitive thing going on between white players." But black and white students, testing the game for *Psychology Today*, rewrote the rules of play . . . shook up the rigidities of the past and introduced free-form alternatives. Black People, though still victims of discrimination, became the agents of change in a game that came to emphasize the absurdities of playing on the same board while living in different worlds.

Every good game has at its heart the magic moment when its structure becomes unimportant and the players take over, caught up in, but not possessed by, the dynamic of the game. If the players of this game—black or white, or better, both—choose to play it freely, they can turn it into "a community spoof of the artificialities of a racist tradition." And once they have done that, they have taken another step toward comprehending—and perhaps changing—the larger game of moving around the playing board of contemporary American society.

NOTES

THE NAME OF THE GAME:

Metro Politics

Order from ——————— Western Behavioral Science Institute
SIMILE II
P.O. Box 1023 - 1150 Silverado
La Jolla, California 92037

Cost ——————— $3.00 sample set; $35.00 for
25 student kit; $50 for 35 student kit

Playing Time——————— 1½ hours (including
follow-up discussion)

Number of Participants ——— 20-50

PURPOSE:

R. Garry Shirts, game architect, describes the purpose: to expose the participants in a vivid and interesting manner to the
problems of the city and some of the political solutions that
have been proposed for these problems . . . The participants, each assigned a role as a citizen of the community, will
have a chance to weigh the strengths and weaknesses of all
the proposals and can form pressure groups to push for passage or defeat of any of the plans (*Metro Politics,* Instructor's
Manual, La Jolla, California: Simile II, 1970, p. 1).

Metro Politics is a highly structured, exciting negotiations
simulation that allows participants to examine six current proposals for metropolitan development. With each player having a specific role definition, plans for solving urban problems
are debated from a variety of differing viewpoints. Equally important to the *issues* involved is the *process* by which decisions are made. For any proposal to gain wholehearted acceptance, a number of coalitions and compromises must be
reached. Discussion and thought about the nature of local
politics is an inevitable bi-product of this simulation.

COMPONENTS:

Metro Politics begins with each player drawing an envelope containing a role description and a specified number of voting chips. The
participants are then free to make public statements about any of the
six proposals for Skelter County and meet with other players. When

three or more people gather together and announce a coalition, the pressure group receives additional voting chips. If the coalition includes certain individuals, known as "good government advocates," the pressure group receives more voting chips.

After twenty minutes, public statements from the various coalitions and individuals are made. Following rebuttals and final re-alignments, a secret ballot is taken of all parties. The proposals receiving the highest number of votes are placed on the final referendum. The chips are returned to their owners and the process repeated, only with fewer proposals. After another twenty minutes of debating, re-aligning, and bargaining, the final referendum is held. Discussion of the results is the necessary conclusion to *Metro Politics*.

THE PLAYERS:

While any number of participants could play *Metro Politics*, the ideal size would be from 20 to 50. Each draws a role description that includes occupation, place of residence, employment, salary, race, and family status. Minority group members must wear identification badges. All players are given influence chips for voting. Although everyone has at least one chip, some have more because they have resources, such as wealth or position. Certain people will draw "good government advocate" cards, which means that they have no special interest to promote except sound government. Coalitions with good government advocates receive extra voting chips.

ENVIRONMENT:

Since *Metro Politics* calls for considerable interaction and movement, a large, open room is needed. A chalkboard and a few desks should be put in the corners and along the sides of the room for use by small groups and coalitions. Ballot boxes should be put up someplace where voters may have privacy. A cloakroom, hallway, or behind a portable blackboard are suitable locations.

The physical setting, however, merely creates the environment for something much more meaningful, namely, an informed discussion at the structure of local government. The relationship between the city and county government has vexed many good government advocates for some time. This simulation sets forth in its proposals some realistic alternatives, the type of urban-suburban conflicts that prevent a better delivery of governmental services.

For students to consider and hopefully advocate municipal government and/or an equitable broad-based tax is bound to make a positive contribution to an informed citizenry. Whatever weakness the simulation has, it compensates for in the content of the game.

NOTES

Theme Two: Life

We are a people that get tired of a thing awful quick and
I believe this continual prosperity will begin to get mo-
notonous with us. We can't go through life just eating
cake all the time. Of course, we like prosperity but we
are having so much of it that we just can't afford it.
—Will Rogers

Typical student life revolves around schooling. Schooling for some means experimental courses, no grades, and a new curriculum. For others it is fifty-minute periods taught by fifty-year-old teachers who use texts from the 1950s. But for all it is a time to define what life is going to be like for the next fifty years.

The institutions of the family and the church, which have been primarily responsible for formulating and filtering the ideas and values that are expressed by youths until they reach junior high school, are eclipsed by their peer group about the time of puberty. The setting for this peer group influence is almost always the school.

With high school comes the press of greater time commitments to extra-curricular activities, and greater external demands for commitments to ideological positions. Coupled with a natural questioning of the values of parents and family, the relationship between the student and his educational institutional setting becomes primary.

The youths' relationship with the church cannot be so easily stereotyped. More and more youths are entering high school without any disciplined religious training. However, an interest in religious questions may have intensified. Also the role of the church as the sole carrier of values may have changed in the minds of many youths, but the search for those values is no less intense.

Personal values, to be effective, need to be formed with awareness of those environments where people are struggling. The concerns must be both Personal and Social. "Out of death comes life." Out of the struggle of our lives comes a witness that reaches into the life space of those who are searching for the meaningful life.

DAVIS, BRODY AND ASSOCIATES, ARCHITECTS
130 East 59th Street, New York City

Robert Rosenwasser & Associates, Structural Engineers
Cosentini Associates, Mechanical Engineers
M. Paul Friedberg & Associates, Landscape Architects

Renewal of this area is being carried out with financial aid from Renewal Assistance Administration, U. S. Department of Housing and Urban Development and the State of New York Division of Housing and Community Renewal.

THE NAME OF THE GAME:

Max's Diner or Intergov

Order from _____ J. J. Mar-Tam and Associates
1053 Delamont Avenue
Schenectady, N.Y. 12307

Cost _____ $25.00 for ditto masters

Playing Time_____ 2 hours

Number of Participants ___ 13-20

The primary game for this chapter is included in its entirety. It may be ordered on ditto masters, or you may want to reproduce copies from this text.

Instructor's Manual

Introduction

Intergov is a simulation game designed to teach the inter-relationship of city, state, and federal government. As a subject, the game deals with the complex problem of low-income housing in the suburbs. Consequently the game can be used either at the end of the study of government or as part of a current problems' course.

Because *Intergov* contains detailed relationships, it is imperative that the participants have a prior knowledge of the workings of city, state, and federal government. Since New Jersey State Government served as the model for that part of the game, *Intergov* is extremely pertinent to New Jersey politics. Teachers from other states may want to alter the titles of three of the participants to meet their states' titling. However, the purpose of the game is not to create a precise mini-government, but rather to teach the relationship between the three branches of government. And this remains constant for nearly any branch of state and federal government and all issues.

Objectives: By the end of the game and discussion, participants should be able verbally to describe the following relationships of city, state, and federal government in the area of housing:

a. State governmental agencies assist in planning and financing, but do not make the ultimate decision.

b. The U.S. Department of Housing and Urban Development can help supply funds, lend planning talent, but does not make unilateral decisions on individual projects.

c. Local citizens may use a variety of methods to postpone or stop actions they consider detrimental to their community. These include political pressure, compromise, and court injunction.

Secondary Objective: Participants will be able to verbally explain the arguments for and against suburban housing of the poor and then choose one of the positions.

52

Assigning Roles

On the day prior to playing, generally describe the set-up of the game. Do *not* tell them the objectives or suggest any action. Emphasize the importance of everyone being in class the following day.

Give each student a list of participants. Then assign each student a role for which he is responsible. The most difficult roles are those of the state and federal government officials (Commissioner of Community Affairs, his assistants, etc.). These parts should be given to students who will look up enough information to portray their characters realistically.

All other parts can be given out at the instructor's discretion. Lively but mediocre students often play excellent leading roles: Godspead, Torie, the Congressman, etc. The reporter should be a quick-thinking, witty extrovert (if one is available). Give people playing Godspead, A. A. Mann, and Rev. Dooley copies of the proposal that they will read at the press conference the next day. They should become familiar with its content and be able to answer questions about it. Instruct them not to show the proposal to other people in the class.

That evening each student should ready himself by (1) reading any government descriptions necessary; (2) thinking about the issue; and (3) defining his role in the game.

Note #1: It is wise to have one or two students prepared to take any role should some vital person be absent.

Note #2: To avoid petty arguments, we strongly suggest that the instructor assign roles rather than call for volunteers.

Directions

Intergov is unique because it allows the game director to participate in the course of the game. The director takes the role of Max, owner of the Unique Diner. He then takes on the task of disseminating information and providing pace for the session. More details are given on the page entitled "Role of the Director."

Step by Step Directions

1. Set up your room into three general areas as designated on the bottom of this page. Included should be (a) a city council chambers with enough seating for all participants; (b) a community church conference room; (c) and the Unique Diner.

The Diner should consist of a long table with chairs or stools to act as a counter and tables or desks pushed together to form booths. There should be either a radio or record player with rock music records. A filled coffee pot or punch bowl adds a finishing touch if either is available.

Two or three conference areas (with two or three chairs pushed together) may be advantageous if space allows.

Identify each of the three major areas with a brightly lettered sign.

Note: Allow yourself plenty of time and three or four assistants to set the room up.

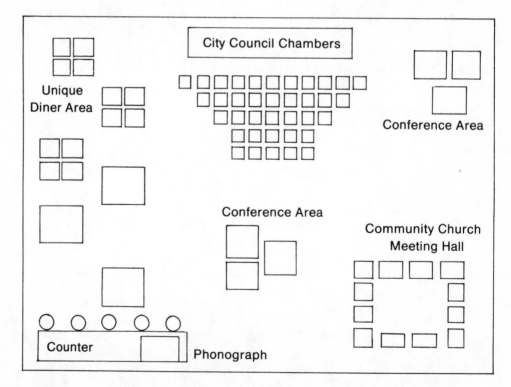

Directions

Amount
of time
allowed

    2. Have the students take a seat in the City Council Chamber.

2 min.   3. Give each student a card and a pin and have him make a name tag (or you may have them drawn up earlier).

3 min.   4. Pass out "Welcome to Blancaville." (Allow time to read each flyer before proceeding.)

0   5. Make sure everyone has his copy of the "Participants" that was distributed the previous class.

10 min.   6. Announce that everyone is at a press conference called by Rev. Dooley, Dr. Godspead, Mr. Mann. Turn the meeting over to them. Pass out the housing project proposal. Dooley should read aloud while others read it silently. Questions follow.

2 min.   7. Game director now becomes Max, owner of the Unique Diner. Explain to everyone that you run the Diner and provide a daily news sheet about Blancaville. Every time they hear loud music playing, they should stop whatever they are doing and come to the Diner for new information. Invite everyone to the Unique Diner.

5 min.   8. Pass out "Blancaville in Brief—June 12." Allow for action and response. Participants may go anywhere in the room.

10 min.   9. Turn up music and call everyone back to the Diner. Pass out "Blancaville in Brief—June 19." Allow for action.

10 min.   10. Turn up music. Call everyone back to Diner. Pass out "Blancaville in Brief—July 8." Allow for action. During this round tell the Mayor that he will chair a city council meeting next round.

20 min.   11. Turn up the music. Call everyone back to the Diner. Pass out "Blancaville in Brief—July 22." Allow spontaneous action for 5 minutes. Then have Mayor call City Council meeting. All should attend.

10 min.   12. Play music again. Pass out "Blancaville in Brief—August 4." Allow for action and meetings. During round tell Albert D'tal that he will chair hearing in

the next round. Also tell Thomas Snupe that HUD will approve funding during the next round.

15-20 min. 13. Turn up the music. Back to the Diner. Pass out "Blancaville in Brief—Sept. 8." Allow spontaneous action for 5 minutes. Then have D'tal call hearing in City Council Chambers.

15 min. maximum 14. Turn up the music. Allow 5 minutes of spontaneous action. Then have mayor call City Council together to decide on housing project.

30 min.

15. Take over and have wrap-up session (questions included in this manual).

2½ hours maximum

The Role of The Director

A. The main job of the director is to keep the game moving. As Max, owner of the Unique Diner, the instructor can pace the game at his own speed. The time frames listed with the instructions are *maximum* guidelines. If a round begins to drag before the time suggested, by all means end that round early and go to the next.

   However, do not extend any rounds beyond the time suggested, except under extreme circumstances. Do not be afraid to end a round while everyone is negotiating and running around. Time is always a factor in life as well.

B. As Max, the director can make quiet suggestions to individual participants *if the game needs new life*. Remember: this is the participants' learning experience! Their own discoveries are vital. Let them determine the action.

C. Run the game with authority. When the round ends, end it conclusively. When a decision is made by the director, he must stick by it. He need not defend it until the wrap-up session. (Like an umpire)

D. During the follow-up discussion, lead the students to make their own conclusion. Simulation gaming is an inductive tool, not deductive. Don't lecture.

E. Move the game quickly and cheerfully. Appear relaxed and confident, even if you are a rookie game director. Running games is easy once the game is under way.

Essential hints

1. *Intergov* should be played in one session with no breaks.

2. Discourage people from leaving the room during the game.

3. Allow no guests or visitors. Everyone should be a participant.

4. Have all handouts stacked in order of distribution. Be sure each handout is given out in sequence.

5. Assign individuals roles opposite their position in life whenever possible. (If you have an integrated class, make sure Martin Torie is black, etc.)

6. Keep a small notepad and pencil handy to write down interesting incidents or questions that occur during the game. Use them during the follow-up session.

7. Do not let the game drag. If one round is going slowly, end it early and begin the next. Keep the pace moving! (Much of this can be done by a brisk, clear voice and eager inflection by the game director.)

Discussion Questions for Wrap-up Session

The wrap-up session is the most important part of the simulation. During this time, the students talk about their roles, discoveries, and questions. Below are a list of questions for the game leader's reference. They are only a guide. The leader's own questions will be more pertinent and situational.

1. Was the outcome realistic?

2. Ask various individuals to explain their feelings as they played their parts (A. A. Mann, Torie, and Schrewd).

3. Ask Commissioner of Community Affairs to explain his feelings. Then ask his staff.
    Note: They should emphasize their powerlessness to make the actual decision.
   Were these feelings realistic? (Get reactions from lesser characters.)

4. Was there any political dilemma for the Mayor or Councilmen?

5. What kinds of things did the state and federal people do?

6. What didn't they do?

7. Who made the decision?

8. Where does political power in a community lie?

9. Discuss any interesting individual roles. Were they realistic? humorous? disruptive?

10. Did anyone feel frustrated during the game? Why?

11. Were you happy with the outcome?

12. What would happen to this issue in this city (or this section of town)?

13. Are there any alternate solutions?

14. What purposes do hearings perform?

15. Are decisions made in formal meetings?

16. How many people changed sides during the game?

WELCOME TO BLANCAVILLE

Situated north of the bustling industrial city of Urbos is the idyllic community of Blancaville. A marked contrast to the smoky, fast-paced city, Blancaville is the ideal retreat from the office grind.

With a population of 12,000, Blancaville offers the ultimate in family living. All residential land is zoned for homes of at least ½ of an acre. The schools are new, progressive, and not overcrowded.

Three shopping centers, one to the east, south and west of Blancaville, provide convenience and quality for modern shoppers. Downtown Urbos is only twenty driving minutes away.

For the future-oriented businessman, Blancaville is a superb investment. Two giant industrial parks flank the town on the north and west. Tasty Soups, RPA, and Ackney Chocolate Company are a few of the industries already located in the area. Within the next ten years, relocation of city-based factories is expected to make the Blancaville shopping areas among the state's busiest.

If you, Mr. Familyman or Mr. Businessman, are looking for a peaceful, quiet, and healthy place, Blancaville should be your new home.

—BLANCAVILLE CHAMBER OF COMMERCE—

Participants

1. The Rev. Thomas Dooley - Pastor of the Blancaville Community Church, Rev. Dooley has emphasized the social mission of the church. More than any other congregation in town, Community Church has attempted to relate to social and political problems.

2. Dr. Amos Godspead - A practicing physican, Dr. Godspead is also a deacon in the Community Church and a leading advocate for the housing project. Not only did he help write the proposal, but he also made contact with key people in the state capital.

3. Mayor of Blancaville, Larry Verbose - Mayor Verbose, considered a "liberal Republican," has been mayor for four years and has led Blancaville to its present position as a hub in the county. He and the council have been very careful not to allow Blancaville to succumb to apartment complexes and housing developers. Blancaville remains a single family, upper middle class town.

4. City Councilman A - see description of mayor. You may accept the role defined by the mayor or choose a more independent route.

5. City Councilman B - liberal mavericks on city council. Vote is decided on each issue by you.

6. Commissioner, State Department of Community Affairs - A political appointee of the Republican governor, the Commissioner has shown some interest in suburban housing for the poor. However, he is subject to political pressure from the governor, legislators, and the public. Some call him a "man on the move."

7. Jacob Stein, Director, Housing Finance Agency - Operating under the Department of Community Affairs, HFA assists individuals and organizations in obtaining funding for housing projects. While with another agency, Mr. Stein worked with Dr. Godspead setting up an urban health clinic.

8. Albert D'tal, Assistant Director, Division of State and Regulation Planning - It is the division's responsibility to work with cities and towns in their technical planning and zoning. Mr. D'tal also answers to the Commissioner of Community Affairs.

9. Martin Torie, President of Blancaville Civic Association - The BCA is devoted to keeping Blancaville from becoming a bustling metropolis. Like most of its members, Martin Torie is a life-long resident of the community. The recent changes in the area have dismayed and angered him. Although the BCA has only 125 members, most are influential citizens in county and state politics. Mr. Torie owns real estate throughout the county.

10. A. A. Mann, Chairman of the Black Unity Conference (BUC), Urbos - College-trained and articulate, Mr. Mann is devoted to the cause of minority development in housing, employment and cultural awareness. The housing proposal is his first major venture. He commands respect from older blacks as well as younger militant blacks in the community.

11. State Senator from the County - defines his own role. Play it as the wind goes.

12. State Representative from Blancaville and another suburb - defines his own role.

13. Newspaper reporter, Walter Lippoff - It is his job to attend all functions, interview key figures, and write a feature about Blancaville for the Urbos *Chronicle*. A resident of Blancaville, he has freedom to write opinions as well as interpret facts. He reads his stories between rounds of the simulation.

14. Field Representative Thomas Snupe, U.S. Department of Housing and Urban Development - When an application for funds for a housing project is received by the HUD office in Washington, a field representative is dispatched to gather further information. HUD has been very interested in suburban housing for the poor. Mr. Snupe should hear all sides and then make a recommendation to HUD.

15. Congressman Jonathan Schrewd (Republican) - A ten-term Representative to Congress from the 6th District, which includes Urbos and Blancaville, Congressman Schrewd has been called a "politician's politician." He can read political winds blowing and usually gets what he wants.

16. Dr. Eugene Thomas - Dr. Thomas is the only black resident of Blancaville. He is a third generation dentist who has been quite active in local Republican politics. The Thomas family owns a beautiful one-half acre estate on Blossom Avenue.

17. Mr. H. B. (Herb) Corpus - A lifelong resident of Blancaville, Mr. Corpus is a leading lawyer in town. As a member of the Rotary, Kiwanis, and Blancaville Civic Association, he has become well-known and respected as a shrewd lawyer and loyal citizen of the town.

18. Citizens of Blancaville (3-5) - As citizens of Blancaville, you may observe, participate, and influence the action in any way. Choose a name and define your own role.

Summary of Housing Proposal

PROPOSAL:
Low income housing development

SUBMITTED BY:
Black Unity Conference (BUC) and
First Community Church of Blancaville

DESCRIPTION:
One hundred units of housing with an emphasis on 3, 4, and 5 bedroom apartments to accommodate large poor families. The project will be built out of concrete blocks and finished with a brick and wood veneer. There will be a playground and community center attached.

LOCATION:
Between 6th and 7th Streets on Central Avenue, property presently owned by the First Community Church.

FINANCING:
The church and BUC will raise and contribute $60,000 for the down-payment. The U.S. Office of Housing and Urban Development will guarantee a bank loan for the balance. The sponsors jointly will pay the mortgage payments each year.

RENT PAYMENT:
Residents will pay rent according to their income. Each rent will be subsidized by HUD funds, which are already established for that purpose.

RATIONALE:
The housing should enable poor people, particularly black people, to move to the suburbs close to decent job opportunities. The housing will also be available for welfare recipients, the elderly, and the disabled, providing they qualify for subsidized rents.

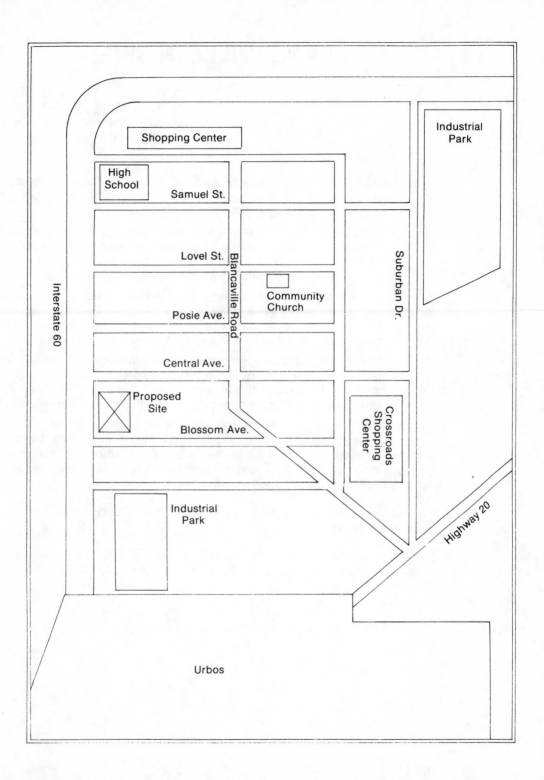

BLANCAVILLE IN BRIEF

June 12

- Courtesy of the Unique Diner—
where the food is always better
than the news.

—NEGROES MOVING TO BLANCAVILLE. The Rev. Thomas Dooley's announcement of a housing project for urban poor people brought the following reactions:

- Republican Mayor Larry Verbose - "This type of project has high ideals but fails to meet the needs of our community. Blancaville is not zoned for apartments of any kind. I see no reason for the zoning to be altered for these people."

- State Commissioner of Community Affairs - "We in the capital are most eager to see how this matter is resolved. All of us are aware of the acute shortage of low income housing. Whether housing in the suburbs is the answer remains to be seen."

- County Republican Chairman Blasam Woods - "I moved to Blancaville because it was a peaceful, safe place for my children. I do not want this beautiful town exploding with people, especially with those who do not fit into this type of community."

- Chairman of the Black Unity Conference, A. A. Mann - "Poor people, black or white, will always be poor as long as they are isolated from good jobs. The factories and offices are no longer in the city. We go where the jobs are. Anyway we have the same right to a decent, clean environment as white folks."

—WELCOME TO TOWN, HEEL-and-TOE SHOE STORE. We hope business will be good.

—MAE BROWN, wife of John Brown, 231 Spruce Street, had triplets at Blancaville Hospital yesterday. TRIPLE CONGRATULATIONS!

BLANCAVILLE IN BRIEF

June 19

- Courtesy of the Unique Diner—
where the food is always better
than the news.

—John Steward, son of Mr. and Mrs. Paul Steward, received his Eagle
Award last night.

—Actress Betty Tart returned to her hometown yesterday. "Isn't it
quaint? Just the way I remember it. Exactly why I went to Holly-
wood."

—The Rev. Thomas Dooley of the Community Church and members of
his congregation will travel to the state capital and Washington to
seek support for their proposed low income housing project.

—Martin Torie has called a meeting of the Blancaville Civic Association
tonight. Purpose: Mobilize support against the housing project.

—Martin Engelthorpe took a two-stroke lead in the Sunset Ridge
Country Club Golf Tournament.

BLANCAVILLE IN BRIEF

July 8

- Courtesy of the Unique Diner—
where the food is always better
than the news.

—State Senator and State Representative to visit Blancaville to meet
with Churchmen and Black Unity Conference Chairman A. A.
Mann. Topic: Housing Project.

—Mayor Larry Verbose announced that he and the city planner will
meet with Dr. Godspead and other citizens to discuss the black
housing project. "We must all listen and discuss the vital issues of
our time," said the mayor.

—Tommy Wheeler of 1216 N. 8th Street won the County Roller Derby.
Next stop—the State Derby in Portwood.

—UNIQUE WISDOM: Once a man gets a handout, the hand will always
be out.

BLANCAVILLE IN BRIEF

August 4

- Courtesy of the Unique Diner—
where the food is always better
than the news.

—Mayor goes to state capital to meet with State Commissioner of Community Affairs who handles state housing. On his drive up, the mayor will probably pass FHA director Jacob Stein, who will meet with our Samaritan, Dr. Godspead.

—Congressman Jonathan Schrewd was again visited by groups for and against the black housing project. According to many, Schrewd can stop the HUD funding—if he wants to.

—Happy Birthday to Sylvia Sunkett, a member of the Board of Education. She is uh, uh, 21 years old today. Isn't it nice to be young again?

—UNIQUE WISDOM: There's no place like home—the way it used to be.

BLANCAVILLE IN BRIEF

September 8

- Courtesy of the Unique Diner—
where the food is always better
than the news.

—HUD approves funds for new housing project. That is, if zoning codes
can be changed.

—Hearing Planned - Albert D'tal, of the State Department of Community
Affairs, will hold a public hearing about the proposed housing
project. "Although we have nothing to say about the decision, we
at the State Department are deeply concerned about this vital
issue. We are holding this hearing so that all voices may be
heard," read his statement.

—How about joining the Volunteer Fire Department? It's your city, too.

—State and national politicians have been asked to attend the hearing
on Friday. Let's make ourselves known.

BLANCAVILLE IN BRIEF

September 21

- Courtesy of the Unique Diner— where the food is always better than the news.

—City Council to decide re-zoning question tonight at its monthly public meeting.

—Bands of Negro youths threw rocks and bottles at police and firemen in Los Angeles last night as firemen battled a blaze in Watts.

—Martin Torie of the Blancaville Civic Association predicts that no housing regulations will be changed. "We do not want people living in rundown housing in Urbos, but we do not want our fair city changed the way the project would change it. What is needed is some kind of creative compromise."

—State Director of Public Assistance said amount of welfare has risen 8% in the past year. More handouts.

THE NAME OF THE GAME:

Powder Kegg

Order from _____ J. J. Mar-Tam and Associates
1053 Delamont Avenue
Schenectady, N.Y. 12307

Cost _____ $25.00

Playing Time_____ 3-4 hours (including debriefing)

Number of Participants ____ 20-40

PURPOSE:

Powder Kegg examines the communications and decision-making process of the American high school. By using a high school typically disrupted by student activists, the game forces participants to play unfamiliar roles and make the decisions for resolving the conflict.

Schools that have experienced student takeovers or other disruptions are likely to find *Powder Kegg* particularly relevant. However, the game is applicable to all standard public high schools.

COMPONENTS:

The solution to any of the numerous problems our high schools face is dependent upon the actions and reactions of at least 5 major systems: administration, established faculty, minority faculty, students, and community residents. This simulation attempts to indicate that each of these components (which are represented by teams, respectively) represents power and must participate in any solution.

The various solutions that are introduced as proposals from each of the teams span a wide spectrum of ways of dealing with the problem of students' rights.

Although many issues cause controversy in the high schools, students' rights seem to be a common thread that links most issues. The way the power systems (five teams) respond to the crisis in Kegg High School gives different proposed solutions.

THE PLAYERS:

Twenty to forty people may play *Powder Kegg* at one time. The participants are divided into five teams—each team representing one of the systems described above. Each group (faculty, students, etc.) is given a goal statement that says something about how it is to play the role. Thus the participants will have to do some decent role-playing.

Each team will choose a chairman and a recorder. On the administration team the leader becomes the Principal; one of the community residents may play the role of the Mayor. Encourage the participants to use their imagination about giving themselves fictitious names and taking on role characteristics. Also do not discourage hyper -or-exaggerated activity; the students may want to initiate a demonstration, and the faculty may feel that they have to picket the school to obtain better salaries.

The game administration will need to borrow one or two of the players to help with the task of running the game. Someone will have to help calculate the points at the end of each round and someone else will have to publish the newspaper.

ENVIRONMENT:
William J. Kegg in Powdertown has had a growing student activism for the past two years. Student protests, walk-outs, and underground papers brought the situation to a head. There seemed no end to the mounting conflict between young and old at Kegg. Finally the State Department of Education offered a large grant to any school that could "adequately involve students in decision-making." The five groups—students, administration, established faculty, minority faculty, and community residents—now have a reason to negotiate.

The first round is spent in team caucuses in which each team chooses two proposals to present to the whole group. This solidifies the team and forces each player to rationalize his position. The players then come to a major meeting where each group reads its proposals. From this point the game follows a set pattern: each round has two parts. During the first part, the players meet in their team caucuses and work on proposals. They may communicate with other teams only by written note (in triplicate). Compromises are reached and a vote is taken. Each group casts a certain number of influence points, depending on its group power, for the proposals of its choice.

The second part of the round permits teams to meet and negotiate. But only two representatives from each team can meet with another team. Thus leadership, trust, and power are focused for half of each round.

The game continues until enough points (70) are accumulated for a proposal or time runs out. *Powder Kegg* not only illuminates the various inadequate means of communication in a school situation, but also permits experimentation with tactics for improving the situation. Both during the game and the follow-up discussion, participants can use their ingenuity for solving the problem.

Decision-making power in schools is often viewed as vested in authority. In many instances, this is a myth. *Powder Kegg* demonstrates how all can participate in a decision. Use of compromise, strategy, and coalitions are essential to the game.

In order to create an environment where all this can happen, a large room with plenty of privacy is needed. *Powder Kegg* demands continuous interaction and consequently produces plenty of noise. Chairs are arranged in five groups, spread far enough to provide each group with some measure of privacy. Two small conference areas are needed in opposite corners.

Essential to the environment are two or three noise producers, usually a record player, a tape recorder, and a radio. These instruments are played loudly and simultaneously during the rounds to produce a high tempo and enough confusion to make the situation realistic. Each round the volume of the noisemakers is amplified as a way of increasing the activity of the participants.

After the game, the de-briefing is held. Allow at least one hour for discussion of the various components of the game. A flip chart or black-board is often helpful in brainstorming for solutions to the problem.

NOTES

THE NAME OF THE GAME:

Consumer

Order from _____ Western Publishing Co.
School and Library Dept.
850 Third Ave.
N.Y., N.Y. 10022

Cost _____ $30.00

Playing Time_____ 2½ hours

Number of Participants _____ 18-36

PURPOSE:

Consumer teaches very simply the principles of buying, borrowing, and saving money. Participants learn how to figure interest rates and evaluate the advantages and disadvantages of borrowing from banks, department stores, and finance companies. The importance of saving is communicated through the experiences of the game and is consequently *discovered* by the players. In a relatively brief period of time, the participants are exposed to the reality of making choices between dull necessities and appealing luxuries. Players must develop sound buying strategies that not only plan for necessities, but also for emergencies. Ultimately *Consumer* is an exercise in individual reasoning and long range planning. Game designer Gerald Zaltman writes, "The major goal of society is the education of adults trained and motivated to act maturely and responsibly. The use of this game, with follow-up discussions, may help the players attain this goal."

COMPONENTS:

The game utilizes three participants in the marketplace: consumer, salesman, and credit agent. The vast majority of participants are given roles as consumers and receive monthly income allotments with which to purchase goods from the salesman. If a consumer desires a product but does not have cash on hand, he must borrow from one of the credit agents.

There are eleven products that consumers may buy. Besides having different prices, each item carries a number of utility points, which are the incentives for consumer purchasing. Each round some of the products change utility point values. Necessities (refrigerator, mattress) have high point values at the beginning but decrease in value with each succeeding round while luxuries (weekend vacations, washing machine) have low initial point values but increase with each round. Other products have constant point values. Consequently consumers are rewarded for purchasing necessities first, and planning ahead for less needed items.

Players may receive credit from three sources: the department stores, the bank, or the finance company. Each credit agent charges a different interest rate and has his own credit terms. For each dollar of finance charged a consumer pays, he loses one utility point. Each consumer must weigh carefully whether it is worth buying on credit at one stage rather than waiting until a later date. The game is designed so that consumers can borrow to buy certain items at the right time and still earn utility points.

If a consumer borrows and cannot meet the payments, his product is confiscated and auctioned. Since each credit agent determines whether to lend money to each individual, participants learn the concept of a credit rating.

Each month every consumer must draw a Chance Card, which represents a surprise event. Some Chance Cards are announcements of sales, others, emergencies. At the end of the game, each player adds up his utility points and subtracts the total interest charges incurred throughout the game. The winner is the consumer who best manages income and credit while at the same time allowing for the unexpected.

THE PLAYERS:

Consumer can be understood by most high school and some junior high students. While there are eighteen roles, as many as thirty-six individuals may play by pairing into 2-man teams. Pairing-up is often advantageous for developing more subtle concepts. Two people playing the same role often stimulate a constructive internal dialogue that is not present otherwise.

The players are assigned roles and given Profile Folders which explain roles and rules. Included are one salesman and three credit agents (two in a small group), while the rest are consumers. As explained earlier, the consumers compete for the highest total of utility points. At the same time, the credit agents compete for the most number of points, which are based on the number of loans minus installments missed and repossessions.

ENVIRONMENT:

A classroom or other large room that allows movement from one playing area to another is suitable. Each credit agent requires his own table or desk. These desks should be spaced so as to allow each privacy for his transaction.

76

The salesman and coordinator should have separate desks, but close enough for them to make necessary transactions. During the follow-up discussion, a blackboard with the point totals for each player would be useful.

NOTES

THE NAME OF THE GAME:

Napoli

Order from _____ Western Behavioral Science Institute
SIMILE II
P.O. Box 1023 - 1150 Silverado
La Jolla, California 92037

Cost _____ $3.00 for a sample student kit
$35.00 for a 25 student kit
$50.00 for a 35 student kit

Playing Time_____ 3-5 hours

Number of Participants _____ 16-40

PURPOSE:

NAPOLI (NAtional POLItics) is designed to teach the operation of the National Congress. The participants go through the simulated exercise of passing or rejecting individual pieces of legislation.

While participating as representatives of the democratic process the students have an opportunity to debate the pros and cons of many of the major issues facing the country today. They must make decisions about limiting the national debt, whether to enlarge the defense budget, whether to seek to eliminate poverty through direct subsidies, and whether or not the United States should be an active participant in the United Nations. These are some examples of the eleven bills that come before the simulated National Congress.

While debating these issues of national concern the "congressmen" must take into consideration public opinion throughout the country with particular emphasis on the districts that they represent. A real purpose is served in this game by students being required to deal with the question of accountability. For example, a "congressman" in the game may feel very strongly that the United Nations is a worthwhile and essential organization but that 60% of the people in his district consider it a part of a Communist plot to take over the country. How does the representative weigh his personal conscience against public opinion which may run contrary to his own beliefs? He/she may rationalize going along with the constituency back home by arguing more diligently for another "good" bill. He/she may decide that the issue is important enough to risk not being re-elected, or the representative may just consider himself/herself as a pawn of the public opinion in his/her district.

In summary the purpose of NAPOLI is to teach participants the working of Congress; however, the secondary purpose is more powerful and perhaps more useful. That secondary purpose is to provide a context where students can debate the great issues of the day and try to determine how to affect the course of human history by means of a representative democratic system.

In addition to the primary and secondary purposes of the game, it performs one other very useful function. It teaches the student parliamentary procedure. In this society where change is dependent upon the enactment of legislation through Congress, anyone interested in being involved in social change must be aware of the parliamentary process that is used in Congress. But more important than the process that is used by 535 congressmen is the knowledge of how to use elementary parliamentary procedure in an organization or even a committee. Very often organizations with the best of intents and best of programs cannot function because they are unable to run an orderly meeting. Knowing how to conduct an orderly meeting as well as participate in a well-run meeting is essential to understanding and participating in real social change in America.

COMPONENTS:

The major components of NAPOLI are two political parties and eight political sub-units which represent the states. In order to simulate the Congress each participant is a member of one of the two political parties (liberal or conservative) and also a representative from one of the eight states.

Each round of the game contains a party caucus and a state caucus. In the party caucus the representatives elect their leaders, and the leader of the majority party becomes speaker of the house. The primary role the party caucuses play is to elect leaders; however, they also attempt to get party unity on some of the major issues. As a matter of fact there is probably more party unity in the game NAPOLI than in the Congress.

The state caucuses represent a time when the representatives of each state (members of both parties) devise a strategy for working together to see that their state receives special projects that will benefit their respective states. The age-old process of "log rolling" or you "scratch my back, I'll scratch your back" is emphasized. A representative's ability to bring projects into his home state is one of the primary factors in determining whether or not that particular representative will be re-elected.

A major component of the political process is missing from NAPOLI. When a civics class goes to Washington to see democracy at work, they often are amazed to sit in the gallery of the United States Senate and see four senators on the floor debating issues of national concern. What students are often unaware of is that the vast majority of the legislative work is done in committees. "Great debates" on the major issues are only a small part of the day-to-day work of the National Congress. Of course, this small part is the most dramatic and easiest to simulate; however, it is misleading to re-enforce a national myth (i.e. most issues are decided through debate in the total Congress).

THE PLAYERS:

The number of participants for NAPOLI can vary greatly. I feel that a minimum of 16 persons is required (at least 2 persons from each state) and probably as many as 40 (5 persons from each state). Fortunately the various segments of each round (party caucus, state caucus and legislative session) move at a quick pace. Therefore, the fact that students are working in relatively large groups is not as inhibiting a factor as it might be.

This simulation game lends itself to having a few students with leadership ability playing dominant roles. Since the only role restriction placed on a participant is that he/she is a member of one of the two parties in one of the 8 states, he/she has a great deal of latitude in how he/she plays the role. Since the work of the Congress usually focuses on a few stellar personalities this aspect of the game is not unrealistic. However, the teacher should point out that this simulation game was not constructed to develop leadership potential.

ENVIRONMENT:

It can be very dramatic and even exciting to turn your classroom into the "halls of Congress" for a day. The party caucuses can take on the air of "smoke filled rooms" and the state caucuses can become strategy sessions for political arm twisting. The dignity, even reverence, that is given to our legislative process is recognized as representative after representative begins his speech with "I rise Mr. Speaker to address the questions raised by the gentlemen from the great state of . . ."

There is, however, one minor flaw in the creation of a good simulated environment. That (Minor) flaw is the names of the states. They are nonsense phrases like agra, coro, efra, baha. These names, like the names of the countries in *Crisis*, jolt the images that the game is trying to create and detract from the natural flow.

In evaluating NAPOLI students who have played the game should be encouraged to make one-on-one comparisons between their game and Congress. How were the bills introduced, politicked through and finally implemented? Such an evaluation will again point out the limited scope of this simulation game. Today most major pieces of legislation are presented by the administration (the executive branch represented by the President). Those not presented as part of the President's legislative program usually have either a strong recommendation pro-or-con from the President or a counter proposal.

Also, the implementation aspect of the bills that are passed is not simulated in this game. If a bill carries a large appropriation it must also become a part of the federal budget; if it is a constitutional amendment it must be passed by a majority of the state legislators.

To point out the limitations of NAPOLI is not to say that it is bad and should not be used; it is merely to remind the operator that an important part of using simulation games is knowing and explaining to the students the limitations of each game.

THE NAME OF THE GAME:

Life Career

Order from _____ Western Publishing Co.
School and Library Dept.
850 Third Avenue
N.Y., N.Y. 10002

Cost _____ $25.00

Playing Time_____ Approximately 6 hours

Number of Participants _____ 15-100

PURPOSE:

The developers of LIFE CAREER ask "how can teachers and counselors help students to understand the society that they will enter when they leave school and to make intelligent plans for their own future lives?" The simulation game LIFE CAREER is an attempt to answer that question by saying that students can actually live a significant number of real life experiences in a simulated environment. As a result of living these experiences they should be better able to make the real choices that face them upon graduation from high school.

In creating the simulated environment LIFE CAREER is designed to meet four basic needs of the high school student. They are: 1. "a feeling for what the future will be like," 2. "accurate information about the alternatives or opportunities available," 3. "a sense of how a life cycle is patterned," 4. "practice in decision making." Each of these ingredients is found in LIFE CAREER and the overall purpose as stated is fulfilled. However, I do want to raise some questions about the implicit value assumptions built into this particular game.

In this game the way to get ahead is to make good grades in high school, go to college, get married later, put off having children, and qualify for a "professional job." The bias that is implicit is that happiness is conditional upon achieving a good education and a good job. The "good life" is tied to the "traditional class values."

This bias is further re-enforced throughout the game by making the game more exciting and helpful to "the bright student." My experience with simulation games has been that the group referred to as "slow learners" in the LIFE CAREER Manual participates as fully (sometimes more fully) than the "bright student." The actual content of the game somewhat contradicts the stated purpose which is:

Perhaps the most significant contribution of LIFE CAREER . . . lies in its potential for clarifying the values and attitudes. Since there are so many alternative career paths in the game and there is more than one way to "win" for any profile simply making game decisions forces the players to consider what is really important for "value" to their person and to examine the value assumptions underlying a given way of life. For example, getting a college education means that one has already made some assumptions about the nature of the "good life." By comparing their decisions—and the consequences of their decisions—with those of the other teams, students may become more aware of both their own values and those of their peer group. Finally, the total game experience may stimulate students to examine critically the values of our society and to reach their own conclusions about the extent to which they wish to conform to or to try to change certain features of our society.

An examination of the underlying value assumptions can take place as a result of this game but it will require skillful leadership on the part of the game operator and/or discussion leader. In my opinion it will mean examining some of the underlying assumptions of the game itself.

COMPONENTS:

According to this game there are four major components in choosing a career. They are: education, occupation, marriage and family life, and leisure activity. As you would expect the occupation that one qualifies for is dependent upon the education achievement of the role person. Likewise the decisions a person makes with regard to marriage and family and use of leisure time affects educational and occupational decisions. The inter-relationship of these four components is realistic in LIFE CAREER and can be helpful to the student who is trying to figure out how to fit together these major aspects of his/her life. In order to make the game more realistic, the component of luck or chance is added into the mix of the other four components. A participant in the game can receive an unexpected promotion or scholarship or can be hit by a serious illness or cutback in employment. However, the chance aspect of the game is somewhat controlled by the other components. For example, if a person waits until he has graduated from college before getting married the chance card on divorce would not apply. This is based on the assumption that persons who marry early are more likely to have their marriage dissolved by divorce. Each of the components relates very neatly to the other components—perhaps too neatly.

THE PLAYERS:

The number of players that can play LIFE CAREER can range from 15 to more than 100. The participants divide into teams of two or three and each group of five teams has a scorer and an application table. The game is very well suited for a class of 20 to 35 persons.

Although it is suggested that LIFE CAREER is designed for high school and college students this game can also be used with junior high students. My suggestion, however, is that it be used primarily with juniors and seniors in high school. It is during this critical period that students are making their "once in a life time choices" regarding education and employment.

ENVIRONMENT:

LIFE CAREER does not motivate the type of internal dynamic that results from a game like GHETTO. The most significant inter-actions take place within the teams rather than among the teams. However, this simulation game can cause an individual student to evaluate some of the implicit assumptions that he/she has been making about the future.

In general LIFE CAREER will create an environment that gives positive re-enforcement to a well-planned career that is developed in an orderly manner. While emphasis is placed on long-term planning (which means that the decisions made in the early stages of one's life greatly affect the options that are available in the future), it is possible for a student to test out different career options. However, very much vacillating with regard to career choice will result in negative re-enforcement in the game context.

The teacher or game operator will have to take the initiative in creating a lively gaming session where students can feel free to experiment with different career options. In some cases this may mean running the game more than once with students playing different roles the second time.

The teacher will have to determine from the group that plays the game how long to run it, whether to re-run the game and in what context the simulation should be used. LIFE CAREER can be successfully used on a one-shot basis (for example, run by the school counselor) as well as an integral part of a total unit devoted to career development. My personal recommendation would be to follow the latter; I believe that LIFE CAREER is more valuable when used with good follow-up material.

Theme Three: Peace

There is no way to peace, peace is the way.
—A. J. Muste

Peace is pacifism; peace is war.

Peace is creative tension; peace is a clear conscience.

Peace is indefinable; peace is what we do not have.

Our inability to achieve it is indeed confounding. In an era where war and the threat of war exist constantly, hate has overcome love, human freedom is denied, communication is dead and life is lost. The absence of personal and world peace may be our most serious human dilemma.

In our culture we have acted upon the paradox that one must wage war in order to achieve peace. The games discussed in this chapter attempt to suggest that there is a better way. The attainment of peace is not only an international problem but also a personal problem, and some of the games will deal with not only achieving peace through international diplomacy but also achieving peace through human understanding.

In addition to the games that appear in this chapter, some games that are discussed in other chapters can also be made applicable to the question of peace. For example, STAR POWER, which is described in the chapter on Freedom, can also be used as a resource for peace.

THE NAME OF THE GAME:

Crisis

Order from ———————— Western Behavioral Science Institute
SIMILE II
P.O. Box 1023 - 1150 Silverado
La Jolla, California 92037

Cost ———————————— $3.00 sample set
$35.00 for a 25 student kit
$50.00 for a 35 student kit

Playing Time———————— 2½ to 3 hours

Number of Participants ——— 18-35

PURPOSE:

The primary purpose of CRISIS is to illustrate how the United Nations works. Although this purpose is never stated directly, it is the underlying principle throughout the game. The participants are divided into six teams, each team representing a nation. Two of the smaller nations are involved in a border dispute centering on the mining of a rare element, dermatium. The national security of the other four nations is at stake in this crisis because dermatium is essential to everyone.

In the Participants Manual, each team's objectives are stated as follows. "1. To obtain the rare element, dermatium, for your people. 2. To avoid destruction of your nation. 3. To maintain yourselves in office." The directions do not state that the participants who are chief decision-makers for each of the six fictional nations should attempt to achieve world peace; however, it soon becomes apparent that none of the objectives can be achieved in a state of war.

Since a delicate balance of power exists with two nations siding with each of the two smaller nations involved in the border dispute, the crisis is negotiated through an international body, known in the game as the "World Organization." Therefore, the more explicit purpose of the game becomes negotiating a solution to an international problem that has no "logical solution."

COMPONENTS:

Although no comprehensive information is given about the six nations in the game, it is apparent that they are not divided into "the good guys" and "the bad guys." Ergosum and Fabuland are the two nations that are involved in the border dispute. Axiom and Burymore are the two nations that have sided with Ergosum, and Camelot and Dolchaveet have sided with Fabuland.

The primary components of the game then become the two power blocks which are vying for ultimate control over the world. The individual nations become sub-components of these two major components so that in addition to experiencing the problem of negotiating among opposing forces, the participants also have the opportunity to participate in negotiation among allied nations. The love-hate relationship and mutual self-interest that hold the friendly nations together in an uneasy coalition often proves more difficult to handle than the negotiations with the opposing power block. The success of the negotiation between the nations and the power block are rated on four "status indicators." At the end of each round, each team receives an indicator which states the probability of that nation becoming involved in a war. The other indicators are "the probability of losing present or future access to dermatium, the probability of losing office by revolution or election, and opinion of international organization of neutral nations about your actions. These ratings provide a somewhat objective evaluation of how each team is playing the game.

ENVIRONMENT:

CRISIS proceeds by rounds and each round represents one year. During the first part of each round, the nations individually make decisions concerning the Ergosum/Fabuland dispute. They may choose to stay out of the dispute entirely; they may attempt to work through the World Organization; they have a choice of sending non-military or military supplies to either side in the dispute; or they may become directly involved themselves, either militarily or non-militarily. The consequences of these decisions are calculated by the game director while the participants convene at the World Organization meeting. The main technique that is used at this meeting is debate. Individual nations may propose solutions which are either voted up or voted down. The World Organization may adopt resolutions chastising or condemning one or both nations in the dispute; they may establish a fact-finding commission; or they may institute a world police force.

At the conclusion of the World Organization meeting, the decisions made by the individual teams are recorded and the round is completed. Playing time for each round is 45 to 55 minutes and playing of three rounds is customary for this game.

A SCENARIO:

My first reaction to the game CRISIS was "it's just not very exciting." As my colleagues and I began to analyze this and other international affairs simulations, we began to conclude that international problems are usually more distant to game participants and therefore are less likely to motivate intense interest. For example, students are likely to be more interested in and knowledgeable about the racial crisis in the United States than about the Middle East. Therefore, when they are involved in a simulation about the relationship between blacks and whites in the United States, their participation and interest is likely to be higher than when they are involved in a simulation about the distant international dispute.

The preceding negative statements are not designed to dissuade you from using international simulations, but merely to alert the game director to the fact that a different dynamic will probably be present in international simulations such as CRISIS.

Partially because of the nature of the simulation, but also reflecting reality, there will be less negotiation among teams in this game than in other games such as *Urban Dynamics* or *Star Power*. National identity, pride and other forms of nationalism do affect the participants and make the problems faced in this game less complex, but more difficult to solve. In the World Organization, the nations are quite likely to be very intransigent in their positions and quite unwilling to break the prescribed alliances in order to form new alliances that might break open new solutions to the problem. The game director should be prepared, in the discussion following the playing of the game, to talk about the various forms of nationalism and their effect upon the way nations relate to one another and negotiate with one another.

Another characteristic of the World Organization is its readiness to send fact-finding squads into the troubled area and its reluctance to take any kind of dramatic action, either positively or negatively. Each nation usually treats the World Organization as necessary but is unwilling to delegate any real power or authority to it.

As a consequence, most of the significant negotiations will go on outside the World Organization and will usually occur in such a way that undermines rather than reinforces the thin layer of trust that exists among the nations.

Although occasionally the World Organization will help the nations move closer to a solution, we have never known a group of participants to adopt a workable solution to the simulated crisis. Often, after three rounds, the crisis is more grave, more complex, and more insoluble than when the game began. CRISIS is written in a way that makes the achievement of a peaceful solution practically impossible. However, the key to solving the international crisis is for the participants/chief decision-makers of the nations to risk the wrath of their native constituency by compromising their position on the issue. This means that the chief decision-makers must take the chance that they will not be re-elected to represent their government in the World Organization. Probably the most powerful lesson of the game is that the attainment of peace is dependent upon a nation's realization that a balance of power is more important than superiority of power.

Although CRISIS is designed to be used when teaching the function of the United Nations, it would also be quite apropos to an examination of the current Middle East dilemma. The primary components of CRISIS correspond to the very real stalemate that exists in the Middle East. Even though there are some distinct contradictions between this simulation and the situation in the Middle East, the basic nature of the problem in CRISIS and in the Middle East is the same.

THE NAME OF THE GAME:

Baldicer

Order from	John Knox Press Box 1176 Richmond, Virginia 23209
Cost	$25.00
Playing Time	3-5 hours
Number of Participants	15-30

PURPOSE:

The name BALDICER means BALanced DIet CERtificate. The object of the game if for each participant to gain baldicers to see that the people he/she represents do not starve.

However, the main purpose of the game is to demonstrate the interdependence of the world economy. In working with this interdependence the players become aware of related issues such as population increase, inflation, technology, and unequal distribution of resources.

The focus, of course, lies with the realization of the effects of the unequal distribution of resources. It is possible for participants to drop out of the action because the people they represent starve to death. Since this is a possibility only for those who are poor one gets a real feeling for the struggle for survival that goes on in many underdeveloped nations. The struggle for peace is seen in the context of an immediate struggle for survival.

COMPONENTS:

The struggle for survival in this game is dependent upon the number of baldicers that each participant can accrue during each round. The factors that influence that process are: work, technology, population growth, natural/social forces, and inflation.

Work in this game is simulated by some nonsense task that the directors demand the participants do in a limited period of time. This "manual labor" can gain each of the participants some baldicers. However, the baldicers can be piled up faster with a food machine or super food machine. These forms of technology multiply the food output and can be purchased only the "haves."

Population growth is self-explanatory and constant in the game. The natural and social forces that operate are included in the form of chance cards. The card a participant draws could read:

  —guerrilla activity and sabotage occurs -3
  —illiteracy eradicated   +5
  —university students riot   -2
  —petroleum discovered   +5

The important factor to consider is that these natural/social forces hardly affect the rich countries, but a couple in a row could spell doom for the "have-nots."

 In the planning period at the end of each round the participants can negotiate for loans, grants, subsidies, or changes in the system. After the transactions during this "planning period," baldicers borrowed or earned and loaned or spent are recorded. If change is to occur in the system it must come during this short planning period.

THE PLAYERS:

 Each participant represents 150 million people. His/her task is to feed these people. As their "food coordinator" he/she must work, attempt to modernize, lend or borrow, and sway the natural and social forces.

 A great deal depends on the individual initiative of the individual participant. Since each participant is for himself, the shy person will suffer during the bargaining. However, there are enough fixed factors built into the game to assure that the purpose does not hinge on "how you play the game."

ENVIRONMENT:

 Baldicer leads the participants through a series of prescribed steps for a number of rounds. Each step corresponds to one of the components discussed above.

 At the end of each round the food coordinator tallies up the number of baldicers he/she has and determines whether or not there will be a chance to modernize in the next round. One may be able to buy a food machine or one may discover that his/her score is 0 or less. In this case, the 150 million people represented by that food coordinator starve.

 After his/her people die, the food coordinator enters an area specified as the "world consciousness." Here the participant can engage in whatever activity seems appropriate to alert the rest of the world to the tragedy that has taken place. The psychological effect of leaving the playing area because your people have starved is quite dramatic. Food coordinators who find themselves "out of the game" often can be helpful in articulating feelings of anger, frustration, and loneliness—feelings that when compounded can create violent conflict.

 On the other side of the coin will be food coordinators who continue to feed their people more and more. The effect of plenty in the midst of famine will have an effect on these participants. In some cases they will use their means to protect what they have; in other cases they will honestly try to help and to change the conditions of their less fortunate brethren. However, they too will experience the frustration of trying to maintain one's own relative strength while helping someone else. Feelings may run high during the playing of the game, so the directors should be prepared to deal with hostility and frustration as well as with the facts of food production.

The feelings of the participants can be channeled during the game into constructive solutions by urging them to view their own self-interest in light of the self-interest of others. A major objective of the game will have been accomplished if the participants can begin to balance their concern for their own survival against the survival of others. Traditionally the "haves" have not been able to view their future in relationship to the future of the "have-nots." Since this angle is only suggested, not built into *Baldicer*, the directors will have to aid the participants in making this connection.

Although this tends to be a "loose simulation," it is one of the best international relations games because it involves and motivates the participants.

NOTES

THE NAME OF THE GAME:

The Money Game

Available in *CONCERN* magazine, March, 1970

Playing Time_____ 3-6 hours

Number of Participants _____ 18

PURPOSE:

The author of the game, Ms. Betty Padgett, consultant to the Church Center for the United Nations, says that the purpose of the game is to simulate some of the economic interactions between "developed" (rich) nations and "undeveloped" (poor) nations.

Peasant revolts, wars of national liberation, and civil uprisings throughout the world are always linked in some way with the economic disparity between the rich and poor nations. In this powerful simulation the participants have the opportunity to experience some of the frustration of trying to develop a balanced system of international trade (the basis for a sound economic system). Because the rich nations have most of the power (money), the task of giving the poorer nations a chance is quite difficult.

The message of this game is that lasting peace is linked with economic security and stability for the entire nation.

COMPONENTS:

Simulated in this game are: the present worth, the economic growth, exports, terms of trade, loans, debt service, grants, and invisibles (transportation, insurance) of seven representative nations.

As the participants move through a prescribed trading cycle, it becomes apparent that the terms of trade, debt service, and invisibles adversely affect the undeveloped countries, while loans and grants are the only factors that work in their favor. Since everyone benefits from some economic growth and exports, the poorer nations must in some way change the terms of trade and loan/debt cycle.

The mechanism that is used in this game to deal with change is a conference fund. Through a process of defining self-interest and negotiating mutual self-interest, the nations produce proposals for uses of the conference fund. There are no restrictions on the use of this fund, so any solution is appropriate and within the rules of the game.

THE PLAYERS:

The game is designed for 18 participants. Two players represent Ghana, Ceylon, Brazil, and Indonesia; three players represent the United States, the Soviet Union, and France; and one person serves as the chairman. Although it is possible to have fewer representatives for the various countries, the figure of 18 participants produces the best internal dynamics for the game.

The chairman can also be the game director. It is his/her responsibility to keep the game moving and to see that the participants are aware of what they should be doing. The chairman will need to have some basic information on each of the countries that is simulated in the game. He/she should also be conversant about general information on international trade.

For the players the game provides a good opportunity to do some advanced research on the countries that they are asked to represent. The players should spend some time finding out about the population, size, per capita income, and exports before playing the game, which provides a context for innovative players to utilize their research.

ENVIRONMENT:

The primary setting for the game is a conference room. Each nation should have its own "working area." Within this environment the game is played in rounds. Unlike most games, each round is different. During the first round the players became acquainted with each other and their own country. The trading components described above are simulated in this round. The second round provides an opportunity for each national delegation to identify its own self-interest and present this to the total group. In the final round the nations attempt to come up with one proposal for the use of the conference fund. This is also the negotiating round and could take as long as two hours to complete.

The playing time for the total game can vary from three to six hours, including the evaluation. An important aspect of this game is that much discussion and evaluation goes on during the playing. However, at the end, it is necessary to pull together some of the proposed solutions and to relate them directly to the reality of international trade.

The game can provide a research model for a class or group to gather data about its own community and test whether some of the same economic factors that create international problems also create domestic problems.

The participants can also follow up on some of the proposed solutions and through research seek to discover where within the framework of the United Nations or another international body, these solutions may have been tried or proposed.

In each case the key to having a significant game experience is informed and creative participants. One can go through the game and learn some of the objective facts about how the rich nations have the advantages in international economics, but real learning takes place when the players get inside some of the problems of the undeveloped countries and begin to struggle with their problems in a creative and informed way.

NOTES

THE NAME OF THE GAME:

Mission VII

Order from —————————— J. J. Mar-Tam and Associates
1053 Delamont Avenue
Schenectady, N.Y. 12307

Cost ——————————— $7.50

Playing Time——————— 1 hour

Number of Participants ——— 25-40

PURPOSE:

By the end of the game, participants will have discovered the effect of imposing one culture on another. *Mission VII* is useful in studying the colonization of Africa, American Slavery, American presence in Viet Nam, and any number of other subjects.

But equally important is the unique type of game design. "While most games available miniaturize and simplify a part of reality or history for the students, *Mission VII* creates an imaginary situation using the space of the classroom. Students are not told the purpose of the activity or the analogy involved. Rather they *discover* the concept being taught" (Steve Greenfield, designer of Mission VII: Instructor's manual for Mission VII, p. I).

Because *Mission VII* is short and not complicated by long instructions, it is a good starting point for instructors who have never led gaming sessions or a difficult class that has never played simulations. Since it requires no prior knowledge of a particular subject, *Mission VII* can be used to begin a new unit to stimulate excitement and interest.

COMPONENTS:

Mission VII plays in one 30 to 40 minute session with no breaks or rounds. The participants are placed in three groups, teams A, B, C. Teams A and B are placed on opposite sides of the room. Team A is given a simplified system of behavior (i.e., move in pairs, governed by male with longest hair) that represents the elements of a culture. At the same time, Team B is also given a system of behavior which is completely different from Team A's. Team C is then given seven tasks (thus

Mission VII to complete in a specific time). Team C is also given the ultimate authority which it can use only as a last resort.

The essential components of any culture—religion, hierarchy, industry, time dimension, dress, and tradition—are reduced to simplified, concrete acts. Each is elementalized for each participant to understand. During the follow-up discussion, each task of the role description can be examined individually and then enlarged to show the elements of culture.

Besides the rational delineation of a culture, *Mission VII* brings out "gut level" reactions to cultural imposition. Members of Teams A and B experience the frustration of being used for another's purposes. They fail to communicate adequately with the colonizers and with the other native group, thus feeling the impotence of being conquered.

Finally groups can be led to understand why it took so long for the native peoples of nations like India, Nigeria, and now South Africa and Rhodesia to form alliances to free themselves from foreign influence.

THE PLAYERS:

Mission VII can be competently played by anyone over the age of thirteen. Of course the inferences become more sophisticated with the maturity of the group. Twenty-five to forty people are divided into three groups, two large groups and one smaller group. The two larger groups, Teams A and B represent two tribes or ethnic groups within a particular geographic area. The smaller group, Team C, symbolizes the intimidating colonizer and exerts ultimate authority in the game.

ENVIRONMENT:

The room used for the simulation should be large enough and secluded enough to allow free and unrestricted movement and voice. A large classroom with some of the desks removed is spacially sufficient. The two large groups should be placed on opposite sides of the room with Team C occupying the large middle. A loud record player with lively music should be playing throughout the game. A flip chart (newsprint or large blackboard) is needed for the game and for the de-briefing session after the game.

Leader's Guide

SUMMARY

Mission VII is a simulation designed to lead students to discover the impact of European colonization on conquered populations. Although it was intended for use in a unit on the Age of Discovery or on 18th Century European History, *Mission VII* may be utilized in the study of missionaries, world cultures, American foreign policy, or a number of other units in which it would be appropriate. Some teachers choose to play the simulation at the end of the unit as the culmination of a study, while others choose it as a lead-off activity to arouse interest in the subject. The instructor's discretion should govern.

If you have played other simulation games, you are familiar with the dynamic and high level of involvement between participants. If you are not familiar with simulation games, playing one is the best way to learn the concept.

Briefly, games are just one more teaching technique among many others. The strength of games is their ability to motivate students to teach one another and also to teach broad concepts.

Mission VII is a somewhat different type of simulation. While most games available today miniaturize and simplify a part of reality or history for the students, *Mission VII* creates an imaginary situation using the space of a classroom. Students are not told the purpose of the activity or the analogy involved. Rather they *discover* the concept being taught.

Even if you have not played or directed other simulation games, you should be able to administer *Mission VII* without difficulty after reading all the game materials and the Instructor's Manual.

The game is designed to be played with 25 to 45 participants and can be completed in one hour. However, if one and a half hours can be allotted, the leader will find the extra time for instructions and follow-up discussion valuable. Students from 13 to 80 may competently play the game.

INSTRUCTIONS

Divide the participants into three teams, A, B, and C. Teams A and B, which represent tribal groups, should have an equal number of members while C, the colonizers, should have half as many as either A or B.

    I. Seat each team in its respective corner and hand out the appropriate role definition to each. (If you have not yet read all participants' handouts, do so now.) Allow three minutes for participants to read their descriptions.

    II. Focus the attention back on the game leader. Next read the description of Team A to the whole group, stopping when needed to clarify the team definition (i.e., point out who has the longest hair and is consequently the team leader). Do the same thing with Team B and Team C. Do not, however, read the list of tasks that Team C must accomplish.

   III. Appoint a leader for Team C. (Just as the King or Queen commissioned officers.) Then allow 3 minutes for teams A and B to begin their roles (i.e., communicating as directed, etc.).

   IV. If there are no questions about role definition from any of the participants, *Mission VII* is ready to begin. (Note: do not define Teams A and B as tribes or Team C as imperialists. Individual discovery comes at different times in the game for each student.) Playing time is 25 minutes. Team C must complete its tasks within this time period. Do not allow more time.

   V. While the game is getting started, the instructor may need to move from group to group encouraging each to follow its description. Team C may require help getting organized to accomplish its tasks. When Team C has a task completed, check it off your list.

   VI. At the end of 25 minutes, stop the game and call all participants into a discussion circle or similar arrangement. Stop the game even if all tasks are not completed. If the tasks are completed before time runs out, merely start the discussion earlier. This is the time for students to share what they learned about their roles, actions, and reactions. The Guide for Discussion which accompanies this manual may be helpful at this time.

HINTS FOR LEADERS

1. Put a few lively participants on each team. But be sure that Team C has a couple of the better leaders.

2. Don't be afraid to put "underachievers" or supposed "slow learners" in leadership positions. They often "dig into" a role with great vigor.

3. Play some loud, lively music during the game. It will add pace and excitement to the play.

4. Have a large stack of paper, pencils, lots of chalk, and a stapler or two available *before* game begins (A MUST).

5. It is often helpful for the instructor to take notes on what occurs during the game. Feeding his information back to the students during the discussion can often fill in participants who did not see a particular event.

6. Stay out of the game as much as possible. Do not give advice or kibitz with participants.

7. Bring a camera to class (videotape if you have access). You are certain to get some good pictures for your bulletin board or newspaper.

GUIDE FOR DISCUSSION

Note: Since each playing of a game produces unique actions and reactions, no list of questions will be sufficient for the resourceful game leader. These questions are only a guide. The leader's own queries and analysis will be more spontaneous and pertinent.

1. Ask each team to tell who they thought they represented in history.

2. Were all tasks completed? How much did Team A and B change?

3. How did you react to being forced into a different working day? (Lead into a discussion of the differing time dimensions of different cultures and the effect of a change on people.)

4. Is any one time dimension better than another?

5. Did you find communication difficult? Why? What remedy is there for two or three peoples speaking different languages in the same geographical area?

6. What did you think of the method of selection of the leader for each team?

7. A member of Team C saying, "I force you . . . ," is the same as either using violence or *threatening* to use violence to accomplish a task. How many times did the C Team use violence? Was this an "easy" colonization?

8. Did any of the members of Team C experience frustration in completing tasks?

9. Does this help explain why colonizers were often brutal and inconsiderate?

10. What was the purpose of the paper slippers? (Shows how foolish one people's customs may be for another people.) The British forced some of their African subjects to wear boots in the jungle. Does this make sense? Likewise, how did the U.S. originally train Vietnamese to fight the Viet Cong?

11. Did any members of Team C feel superior to the other two teams?

12. What feelings did members of Teams A and B have toward Team C?

13. In conclusion, how do you think African or Asian peoples reacted to the colonizers who took over their lands?

TEAM A—ROLE DEFINITION

As a member of Team A, you are to do the following things:

1. Stay in your corner of the room unless moved by a member of Team C.

2. Move very slowly at all times.

3. Spend your time drawing pictures of animals.

4. Sit, stand, and move in pairs.

5. Communicate by speaking to other members of your team. You cannot speak to members of the other teams.

6. Accept orders from the male with the longest hair. He is your team leader.

7. Worship the symbol on the wall in your team corner.

8. Have no contact with Team B.

Your goal is to stay just as you are described. You do not have to do anything else unless a member of Team C says "I *force* you to do this task. . . ." Then you must do as he directs.

TEAM B—ROLE DEFINITION

As a member of Team B, you are to do the following things:

1. Stay in your corner of the room unless moved by a member of Team C.

2. Work for only two minutes, then rest for two minutes, work for two minutes, and so on.

3. Spend your time copying a page out of a book.

4. Sit, stand, and move in groups of three.

5. Communicate with each other only by writing. You may *not* speak.

6. Accept orders and instructions from a brown-eyed girl who is elected by a vote of all brown-haired girls. She is your leader.

7. Worship the symbol on the wall in your corner of the room.

8. Have no contact with Team A.

Your goal is to stay just as you are described. You do not have to do anything else unless a member of Team C says "I *force* you to do this task. . . ." Then you must do as he directs.

TEAM C—ROLE DEFINITION

As a member of Team C, it is your goal to have the following missions completed in twenty-five minutes.

1. You should try to complete each task with as little resistance or objection as possible. In other words, try not to arouse anger from the other teams.

2. If, however, members of Team A or Team B refuse to do as you ask, say "I *force* you to do this task. . . ." They must obey. If any still refuse, report them to the game leader.

3. Try to complete the tasks as quickly and completely as you can.

4. You can communicate to Teams A and B only by sign language, drawing pictures, and gesturing. *No* talking.

5. As soon as you complete a task, tell your instructor so he can check it off your list.

TASKS

1. Move 6 people from each team (A & B) to the middle of the room and have them sit down.

2. All the males on Teams A and B must draw twelve figures of people on the blackboard in one minute's time. Then erase the board and have them do it again.

3. All women on each team must tear paper into human shapes. They should work for 5 minutes, then rest for 30 seconds, and then repeat the activity. Make sure all work at the same rate or speed.

4. You must move people in groups of four. Force Teams A and B to work together.

5. Five people from each team (A & B) must make a pair of slippers out of paper and wear them around the room.

6. Anyone with blue eyes is to be put in a separate place—called a school—where they do no work but worship a new symbol.

7. All brunettes (people with dark hair) should do double work.

NOTES

DISARMAMENT

Leader's Guide
A simulation dealing with trust, negotiation, and conflict.
Adapted from an intergroup exercise created by
Norman Berkowitz and Harvey Hornstein.

PURPOSES:

To feel oneself expressing and utilizing trust and to tag feelings about it.
To recognize factors contributing to our preventing trust.
To identify tactical alternatives.

SCHEDULE:

A. Introduction and explanation of the game.
B. Game One—A game consists of a series of rounds, up to 10 per game. Although negotiations may be called for after each round (if both groups agree), the negotiators *must* meet after rounds 3, 6, and 9.
C. Game Two
D. Reflection on the simulation, sharing of learnings. As in all behavioral simulations, the reflection period is crucial to the learning process. No simulation should be used unless reflection is included.

SETTING:

Three rooms are needed. This is a simulation for two groups, each occupying a separate room. Negotiations must be held outside of the group rooms.

OBJECTIVE:

Your primary objective is to end the exercise with as much money as possible in your group's treasury. Your objective is not necessarily to end the simulation with a treasury larger than that of the other group, but to enlarge your own treasury as much as you can.

TO BEGIN:

1. The two groups should go to their separate rooms.
2. Select a treasurer, a negotiator, and a recorder.
3. Each group should collect $2.00 per member for this simulation.
4. 60% of this contribution remains in the group as its treasury.
5. 40% is placed in the Common Bank, managed by the umpire.

TO CONTINUE:

6. The umpire will give each group 20 cards. Each card has a marked and an unmarked side.
7. Place all 20 cards on the floor, marked side up. The cards

should all be visible at once.

8. The simulation will proceed with three (3) minute rounds. The umpire will signal the beginning and the end of each 3-minute round.

9. In each round your group must decide to turn 0, 1, or 2 cards. This process will continue round by round until the game ends. It is possible for a group, if the game goes 10 full rounds, to turn all of its cards.

10. For reasons which will be clear later, a group may choose to turn a card which has been turned on a previous round, returning it to its original state. Of course this can only be done as part of a round by your group.

GAME ENDING:

A game may end in either of two ways:
(1) Completion of 10 rounds.
(2) One group calls for an "Attack."

PAY-OFFS:

(1) The pay-offs are related to the way in which the game ends. If all 10 rounds are completed:

a. Your treasury will be paid by the Common Bank 2¢ per member for each blank card showing.

b. Your treasury will pay to the Common Bank 2¢ per group member for each marked card showing.

c. Thus if all 20 cards are blank after 10 moves the group treasury gains 40¢ per group member from the Common Bank.

d. If all 20 cards are marked after 10 rounds, the group treasury loses 40¢ per member to the Common Bank.

(If the game ends with all 10 rounds completed, it is better to have blank cards than marked cards.)

(2) If one group calls for an "Attack":

a. The umpire notes the number of marked cards showing.

b. The group with more marked cards showing gains money for its treasury, while the group with fewer marked cards showing loses money from its treasury.

c. The amount of money changing hands is determined by *how many more* marked cards are held by one group as compared to the other.

d. The group holding more marked cards receives from the treasury of the other group 4¢ per group member for each marked card *above the number held by the other group.*

e. If both groups have the same number of marked cards showing, no money changes hands.

f. An "Attack" can be called by a group by informing the umpire of this desire as part of your move during a round of play.

(If a game ends with a call for an "Attack" it is financially better to have marked cards showing.)

NEGOTIATIONS:

Between rounds you will have the opportunity to communicate with the other group through negotiators.

    a. For this purpose, each group must select a negotiator. You may change your negotiator at any time.

    b. These negotiators must meet after rounds 3, 6, and 9. Negotiation sessions will last 3 minutes.

    c. Optional negotiations can be called for after any round, *if the other group agrees.*

    d. Negotiations will be held outside of the group rooms.

    e. Negotiators may choose to be completely truthful in their discussions, but they are not obliged by the rules to be so.

    f. Negotiators may choose to be completely truthful in their reports to their groups, but this is not required.

UMPIRE:

The umpire will coordinate the activities of the two groups and manage the money in the Common Bank.

RECORDER:

Each group needs a recorder. His task is to keep a running record of the group's decisions and moves round by round. This report serves as a kind of "group memory" during the reflection period.

TREASURER:

The group treasurer is to manage the group's treasury. He must make all payments required at the conclusion of the game.

SOME REMINDERS:

1. At each round you may turn zero, one, or two cards from one side to the other.
2. You have 3 minutes to decide on your move each round.
3. If the game ends in 10 moves, it is financially wise to have blank sides showing.
4. If the game ends with a call for an "Attack" it is wise to have marked sides showing.
5. During each game, only the negotiators may leave the group rooms.

DISCUSSION GUIDE FOR "DISARMAMENT"

A. Questions dealing with the way the game was played:
1. What happened in the game? Who did what to whom? Which round? (Note: It often helps the group if the game leader places on the chalkboard the round-by-round behavior of each team.)
2. What was the communication pattern in the game?
3. Was there a turning point in the game for your team? If so, when and why?
4. How do you feel about the way your team played?
5. What do you think you should have done differently?
6. What do you wish the other team had done differently?

B. Questions dealing with intra-group problems:
1. How did decisions get made in your group?
2. Who influenced whom? How do you account for this influence pattern?
3. Who exerted most power in policy formulation? How was this power exerted?
4. Who exerted least power in policy formulation? What form did this abdication take?
5. Who disagreed with policy? How was this expressed?
6. How did the team deal with dissent? Did a standard get set that forced conformity? Did dissenters feel that they had to "go along with the group"? Or did the team permit open dissent? (Are we talking about "loyalty" and "patriotism"?)
7. Was there ever a revolt in your team? Did any people attempt to take over the leadership? What form did this revolution take?
8. Did any team members "die" during the game? What happened in the team to bring about such feelings of apathy? Power-lessness?
9. Was the money given equally? Or did one or two people provide the capital for the others? If so, what effect did any financial imbalance have on the way your team operated? Were capitalists more influential than others?

C. Questions dealing with inter-group problems:
1. To what extent did either team project on the other its own worst or best motives? How do you account for this stereotyping? Is there an ethical issue here? Did either team de-personalize the other team? See it as "the enemy," Dinks, Gooks, etc.?
2. To what extent did either or both teams misperceive one another's motives and behavior? What happened? What led to this? Was it corrected before it led to disaster?
3. What rationale was agreed upon by each team to justify its behavior? If one team attacked another, how was the attack justified? Does this rationale bear scrutiny? Was there sufficient reason?

4. Did either team indulge in propaganda about itself? If so, how and when?
5. What led to increased fear or increased trust between the teams?
6. What was the role of the negotiators? Did either negotiator tell the truth? Tell lies? How was the latter behavior defended?
7. To what extent was your team influenced by the money you invested in the game? How were you influenced?
8. To what extent did each team buy into the "win/lose" norms of the game? How do you account for either the high competitiveness or the high collaboration and trust that occurred?
9. What efforts did either team make to increase trust? Did you:
   a. Exchange hostages?
   b. Spy?
   c. Have on-site inspection?
   d. Visit each other's rooms?
   e. Exchange your cards?
   f. Exchange money?
   g. Agree upon a common enemy, such as the World Bank?
10. If there was an attack, when was this policy decision made? What instructions did you give to your negotiator?

D. Questions based on ethical reflection:
   1. Did either team challenge the game's norms? Based as it is on a win/lose context, each team had the option of being competitive or collaborative. Did either team challenge the game's parameters and operate "outside the law"? Civil disobedience: was there any? What forms did it take? What was the response?
   2. In this game, what does "justice" mean? How could justice have been promoted? Does justice always stem from the winner or the group in power?
   3. What does "liberation" mean in the context of this game? How could each team have liberated the other from fear of attack? Was this ever a goal? What is the ethical issue here? How does liberation relate to the New Testament?
   4. What does "reconciliation" mean in the context of this game? What factors led to reconciliation? What factors destroyed "Community trust"? Since reconciliation is a New Testament term and a goal for Christians, to what extent did you in this game seek it? Why or why not? Where is the ethical issue here? Did either attempt to convince the other of its guilt (war crimes) and seek repentance from the other before forgiveness and reconciliation could take place?
   5. How was power used in this game? It is said, "Power does not corrupt, it reveals!" What was revealed about our human situation in the ways the different teams used their power? What ethical decisions did each team have to make about its use of power? What were the alternative uses available?
   6. What part did speaking truth have in the way you played? Were there degrees of truth spoken or implied? What is the ethical issue here?

7. Is there an ethical issue around the gathering of intelligence (spying)? What is it? What limits were set on spying by the teams? Why?
8. **What ethical method was utilized by each team:**
a. Ethic of ends (idealist ethics)?
b. Ethic of duty (legalistic ethics)?
c. Ethics of responsibility (response ethics)?
9. Did either group make the Sermon on the Mount a law? What was the result? Did this legalism make life easier for the team insofar as it no longer needed to revise its policy round by round?
10. Was the "demonic" at work in the game? For instance, did an honest mistake by one team result in disaster in any way?

E. Questions dealing with international affairs:
1. What parallels do you see between this simulation and world events?
2. How does the simulation reveal some of the problems facing the United States and the USSR in their disarmament negotiations?
3. Can you relate this simulation to experience of the International Disarmament Commission?
4. Does the simulation give you any insights into the complexities of disarming? The place of trust? The place of fear? The place of misperception?
5. Does the simulation frighten you when you relate it to the dangers of accidental nuclear war? Of policy based on incorrect assumptions about the motives of the other nation?
6. Does the simulation remind you of the role played by ideology in international affairs? Who is ideological? Where and when?
7. Can you outline briefly the international situation about armaments? What are the facts?
8. Can you identify international situations where spying helped prevent warfare? Spy planes over Cuba? Russia? China? USA by Russia? CIA? Other?
9. The patriotism issue is often raised to prevent non-conformity in public opinion. Can you see parallels between this game and the situation in your own nation?
10. Are there ways the nations might set new norms, outside the traditional win/lose norms that often operate, so that their behavior promotes trust?
11. Do nations ever lie to each other? To their own people? Did your negotiators do anything like this to you or each other?
12. Was there civil disobedience in the game? What effect did it have? Is there civil disobedience in the nation over foreign policy? What effect does it have?

NOTES

Theme Four: Love

—love is buying someone a present
with your own money
—Charlie Brown

If you were to ask me, "Do you love your wife and son, the members of your immediate family?" my response would be quick and affirmative. Of course, my wife and I occasionally disagree on some things, and my son can really irritate me at times, but never do these momentary disagreements or irritations cause me to doubt that I have a deep and abiding love for them.

If you were to move one step further and ask me, "Do you love your neighbor?" I would hesitate just a minute and ask myself, "which neighbor?" Of course, I love the neighbor down the street whose children play well with our son, but I am not so sure about the neighbors just up the block who have the wild all-night parties. But wait, you are talking about love in the biblical sense, and I am responding from a personal perspective. OK, you are right, there should be no distinction. I should love everybody, because everyone is my neighbor. To love them means to care for them and to seek to develop a meaningful relationship with them.

I can be *careful* to maintain a relationship with my neighbors with whom I am in contact, but what about the neighbors on the other side of town, in another state or country? My guess is that to be neighborly with those I do not know means trying to learn more about that neighbor. If I want to know more about my neighbors in Bolivia, my quest may some day lead me to express an overt act of love by setting up a student exchange program that will bring me and my neighbors to the deep, deep South in direct contact.

A third question that you should ask me in order to push me on my definition of love would be, "Do you love your enemies?" Now that is a hard one, because I have political enemies, who I am trying to defeat, and social enemies who I am trying to ignore. When Jesus said, "Love your enemies," I can see what he meant in the abstract—the ideal is something that we must keep in mind. However, when taken in the concrete, "Love your enemies," presents problems. If I think about loving the person I am contending with, I may lose.

Perhaps we can link the command to "Love your enemies," with the command to "Love your neighbors" and come up with an answer. When we play a game like *Monopoly* (which will be described below) the opponents are at once the enemy and the neighbor. My feeling is that an enemy is just another neighbor. However, I must work a lot harder to know and to care for the neighbor who for the moment is considered the enemy.

For decades *Monopoly* has been one of the most popular games of competition. We have always had fun and learned quite a bit about real estate by trying to win over the other players. My bet is that you can have just as much fun and also learn something profound about love by playing *Monopoly* in a way that makes everyone the winner. We may have to work harder at the game. However, I am convinced that getting to know the problems of one's opponent and learning to care enough to make these problems one's own opens the door to having someone else experience my problems and come to care about them as his own. This mutual understanding and caring is the experience of love.

THE NAME OF THE GAME:

They Shoot Marbles, Don't They?

Order from ——————— Urbex Affiliates Inc.
474 Thurston Road
Rochester, N.Y. 14619

Cost $40.00

Playing Time——————— 1½ to 3 or 4 hours
(discussion time included)

Number of Participants ——— 8-40

PURPOSE:

Designed by Dr. Fred Goodman of the University of Michigan, MARBLES was originally intended to provide an experience of the dynamics and difficulties in police-community relations. In this capacity, it has been used successfully to help police and community groups explore what could be done to bring about a situation more conducive to a better relationship.

But MARBLES has been repeatedly modified, both at Michigan and elsewhere, and now offers a way of exploring how varied groups can best pursue their own interest and at the same time interact with each other in just about any context—from the society-at-large to a specific institution or group. Further, it provides for repeated experimentation for testing out a variety of approaches in an attempt to find the most workable way to make the society or the group function to the greater good of each part.

COMPONENTS:

It is the game's fluid structure which allows for unlimited adaptation to any situation. A few marble shooters sit around a cardtable and shoot marbles, following a set of procedures which quickly direct all their interest into trying to earn a maximum number of marbles either individually or as a group. All other players are divided into teams representing rule-makers (administrators), rule-enforcers, a judge, an opposition party (to the rule-makers), and perhaps property owners, land owners, lawyers, a banker, social workers and any other team with a role important to the total system.

The game operator, called the Game Overall Director (G.O.D.) keeps in play all those forces which people must conform to but cannot make decisions about—a kind of body of natural law. Basic subsistence must be secured. (Every player must turn over one marble per round to the G.O.D.) Deals between persons which depend on negotiation can last as long as negotiation is not broken down. Rule-makers must periodically win the right to continue in that capacity. The task of the rule-enforcers is to find a way to enforce the rules, etc. All other "rules" governing play are subject to the decisions or whims of the players and the kind of interaction which results once a variety of self-interests are being pursued simultaneously.

ENVIRONMENT:

MARBLES is among the most instantly involving of simulations. Everyone is given one or more marbles to start with, and, almost immediately—certainly before the instructions are complete—begins pursuing the goals of his team. What the goals of the total group are or should be usually are not clear to anyone, nor of very great concern either.

After the game has gone on an hour or so, it is best to stop it for discussion. Players usually begin to critique each other's—and their own playing—and then ask to try again to see if something better can be worked out. The game is infinitely flexible; new approaches to creating a human and functioning system can be tried as long as the players want to continue the attempt.

The charged atmosphere involves everyone in the pursuit of a just society where the needs and concerns of all are met. But the pursuit is ever-elusive. One method might work better than another, but the perfect formula for people to interact together for each other's good remains out of reach.

The game has not been boxed, since its materials are inexpensive and can be easily gathered together. Collecting them is well worth the effort, for the game provides a forum for the discussion of what are at once the simplest and yet the most complex factors of human interaction.

122

NOTES

Parlor Games

Loving and being loved is not only everyone "winning," but also cooperating to develop ways of achieving the goal of winning without others losing. In other words, one way of loving one another is writing new games that have the goal of making everyone the winner. Another approach is to rewrite existing parlor games so that everyone wins.

A leader can rewrite any of these games (*Monopoly*, *Clue*, etc.) and have his group play them. But far more creative and significant to finding alternatives in real life is having one's group rewrite the games to eliminate interplayer competition. The act of rewriting can be far more satisfying and enjoyable than the playing of the rewritten game.

Holding rewriting sessions is a good way to use draggy time periods before vacations, slow weeks, or split classes that normally would be wasted. Small group projects as a part of units on Modern Culture or the Future are other possibilities for rewriting. Intellectual and emotional interaction are stimulated by rewriting. Here is how a 5-day workshop could work:

Day 1
1. Divide your group into sub-groups of 3 or 4. Have a third of them bring in a game the following day.
2. Discuss the purpose of parlor games. Make sure that concrete examples of purpose, means of winning, and scoring methods are brought up. Ask for suggestions on changing games from "win - lose" to "win - win" models.
3. Assign them the task of working on a game that night.

Day 2
1. Groups form and rewrite a game from "win - lose" to "win - win."
2. Have them begin to play the game. Assignment: think of any changes needed.

Day 3
1. Allow ten minutes for revisions if necessary.
2. Play the game for the rest of the period.

Day 4
1. One person stays with each game while all others choose another game to play.
2. Play the game for the rest of the period.
3. Assignment: write review of new rewritten game.

Day 5
1. Discuss the implications of changing purpose of games. Participants should dip into their emotions, cultural influences, and new view of parlor games. Keep this part of the discussion on "feelings;" stay away from game designs and specific games. This is a time for determining whether it is possible and advantageous for everyone to win. It is fun if nobody loses.

2. Go over each game individually paying particular attention to the new purpose. Have each group briefly explain the game and then have other players critique it. Make any other necessary changes and file the game for future playing.

While some games do not adapt to new purposes, most parlor games can be rewritten by almost anyone. Here are some examples of how to rewrite games.

Beat the Clock (Milton Bradley, $3.95). Based on the former television show, *Beat the Clock* is a simple, active game in which a group divides into teams and competes against each other through a series of physical stunts. The method of winning, not surprisingly, is completing the task in less time than your opponent team. One way to revise this game is by abolishing the team system and making the clock your sole opponent. This will make *everyone* root for the person doing the stunt. An even better way would be to make up your own stunts so that each one includes all participants in the game. Set a time limit goal (2 minutes to do such and such) and see if the group can reach the goal. The more people that cooperate, the better chance they have of beating the clock.

Scrabble is a well-known word game in which every player gets ten pieces of wood, each bearing a letter and a point value. The players move in turn, building words on a board and accumulating points. Ultimately the board is an intricate design of words interrelated to each other. The person who has the highest number of points at the end of the game is the winner. *Scrabble* is a gold mine of possibilities for revisions. For example, set a point value goal for the whole game (by the end of the game, we will have built 8,000 points worth of words) and work together to reach the goal. Or, if players still want to come up with a winner, award a specific point bonus every time one player gives a letter to another player. By cooperation both players win.

Bonuses for key words (*cooperate*, *peace*, etc.) when two or more players build them together is another way to encourage working together. Once the game is underway, dozens of possibilities for rule changes spring up. Every game—just as every life—has viable alternatives to the present limitations.

Executive Decision (3M Company, $8.95). Just as the name indicates, *Executive Decision* deals with corporate decision-making. Each player is a head of a company that is part of the competitive marketplace. Purchasing raw materials, producing, setting prices, and selling are the tricky elements of the game. Obviously the winner is the player who amasses the most money.

Again, there are many possible revisions. The most dramatic would be to change the purpose to everyone's earning X number of dollars. More subtly, one could set profit limits, compensatory payments for low achievers, or coordinate purchasing. Each group of players would undoubtedly create different rules. One thing that could be tried is to have three different groups rewrite and play *Executive Decision* and write up the consequences of their changes. Then have the groups get together and discuss all the changes made.

NOTES

THE NAME OF THE GAME:

Monopoly

Order from _____ Department Stores, Bookstores, etc.

Cost _____ $2.95 to $6.95

Playing Time_____ 2-6 hours

Number of Participants ____ 2-8

PURPOSE:

Monopoly, perhaps the most popular of all simulation games, has seldom been thought of as an educational tool. This game, developed by Parker Brothers, is used primarily for entertainment purposes. However, it actually teaches the participant a great deal about buying and selling real estate and about our economic system in general.

The economic value of ownership is probably the single most important thing to be learned from Monopoly. To own "boardwalk" or "park place" will almost assure victory at some point in the marathon gaming process. But ownership of blocks of the less valuable property also proves very beneficial. In short, the only way to survive in this simulated economic environment is to own property.

A second major purpose of Monopoly is to teach the value of competition. Striving for money is the "carrot" that propels the player to "work" harder to beat his colleagues. The desire to win—to wind up with more money than the other players—is the object of the game.

Finally, one must admit that Monopoly was designed to create an atmosphere where people can have fun. The fact that Monopoly has accomplished this objective is easily measured by tallying the sales of the game over the nearly forty years of its existence. Monopoly is not merely a game of chance where the one who rolls the right numbers wins; it is also a game of skill and those who develop the skill have a great deal of fun.

COMPONENTS:

Monopoly contains most of the major components of our American economic system: wages, taxes, property ownership and development, philanthropy, utilities, transportation, law enforcement, and corporate interests in the form of the bank.

Although none of these components conform to the manner that they actually function in our everyday life, the game does simulate enough of reality to be an adequate teaching tool. For example, the same person in the game buys land for himself and rents periodically (depending on the roll of the dice) from other players. Taxes are charged only when one is unfortunate enough to land on the wrong square or draw the wrong card. But one does soon learn that the system is designed to benefit the owner and that taxes are figured on the basis of your net worth financially.

To briefly recap how each of these components fits into the game, I shall merely list their respective functions:

Wages: Each time a player passes "go" he/she receives $200.00. This is an indication of the game's compromising with reality because every person in our society does not make the same wages.

Taxes: The board reserves two spots for taxes. If a player lands on the spot marked "income tax" or "luxury tax" he/she pays accordingly.

Buying and selling property: A player must own a block of real estate before he/she can develop it (i.e., build houses and hotels). The bargaining and bartering that mark the real estate market come alive in *Monopoly*. The factor that is out of step with the present real estate system is that today almost everyone buys property by obtaining a mortgage. In *Monopoly* mortgages are much less significant.

Philanthropy: An entire batch of chance cards in *Monopoly* is referred to as "community chest" cards. By drawing one of these cards, when one lands on the appropriate stop on the board, the player either "gives" or "receives" according to the directions.

Utilities: The "electric company" and the "water works" represent our system of private ownership of public utilities. Rates in the game are dependent upon a roll of the dice; one sometimes wonders how this differs from the way rates are set today. Also not unlike our present set-up, a monopoly of the utilities means higher rates.

Transportation: The four "railroads" are the only form of transportation, but they also accurately illustrate how private ownership (particularly a monopoly) affects a public commodity. Although it is inaccurate to say that fares quadruple when one person owns all four railroads, it is fair to say the return on a person's investment does increase significantly as he/she moves toward a monopoly on the railroads.

Law enforcement: Although it is a very weak component of the game, the jail does provide the reminder that one has to contend with the police. Going to jail is based on the roll of the dice.

Corporate and government interests/the bank: Sliding interest rates, government controls, and political considerations are not a part of *Monopoly*. However, in the game the bank never goes broke. The lack of corporate interests prevents *Monopoly* from being a realistic economic simulation, but if one thinks of the bank as a representation of all those absent outside controls, the individual buying and selling takes on educational significance.

THE PLAYERS:

Monopoly is designed for 2 to 8 participants. My feeling is "the more the merrier." So much of the fun and excitement of the game is in

130

the interaction around the board. There is enough activity to keep eight participants busy between turns.

Another strength of *Monopoly* is that it is of interest to almost everyone, from elementary school to adults. It is a natural for peers or for the family.

ENVIRONMENT:

Monopoly can be played in different environments. The most obvious is the fun-and-games situation: A family, a group of friends, or some acquaintances gather for a good time.

For our purposes, however, we can examine two "illogical" environments. First is the teaching situation. As was indicated above, *Monopoly* can teach quite a bit about the American economic system. Can you imagine taking class time to play *Monopoly*? My guess is that much can be learned by taking a popular game like *Monopoly*—a game that is not generally considered a teaching game—and convert it into an educational tool.

A second environment can be created by changing the objectives of the game. One normally plays *Monopoly* to win—to get more money than the other players. Why not play to tie—is it possible to play *Monopoly* to wind up with the same amount of money as the other players? Why not?

In the second environment which will be discussed further below, the emphasis is on cooperation, rather than competition. This type of game setting has psychological, as well as economic overtones.

A SCENARIO:

Imagine that you are with six friends, camping in Nova Scotia. You are settled in for the week and after two days discover that the only form of evening entertainment is the *Monopoly* game that one of your friends brought.

At 7 o'clock you begin the proverbial, "winner take all" *Monopoly* game. By 9:30 the jovial atmosphere has become a little tense. Three of the seven players have obtained monopolies and it is only a matter of time for the other four. They know that the die is cast and they begin to resent their second class economic status. But the smell of victory is in the nostrils of the players who are winning. They ignore or make fun of suggestions to quit. They are spurred on by the sense of competition among the "top three" and by the sweet feeling of superiority over the more unfortunate participants.

Finally one of the participants is able to halt the action long enough to point out the parallel with modern day society. Those who "have" keep winning, while those who are "have-nots" keep losing. The result is bitterness on the part of the "have-nots" and a keener sense of competition on the part of the "haves." So the suggestion is made that the players begin to try to correct the injustices in their present game of *Monopoly* to see if they can get any clues as to how to cure the injustices of the economic system in America.

All the players now agree that the objective of the game shall be to redistribute the wealth evenly rather than to see who can win all the money. The first rule change is that all persons who pass "go," who have less than five hundred dollars, receive four hundred dollars rather than two hundred. Those who have more than five hundred dollars receive nothing. From this point you begin to experiment with rent control, a graduated income tax, and the mandatory break-up of monopolies. Everything that you try has some success, but the right combination is hard to find.

It is now midnight. You have not successfully redistributed the wealth, but no one wants to quit. The whole mood of the players has changed. Before creating the new rules hard feelings were beginning to develop between the "winners" and the "losers." Now a new sense of cooperation has permeated the atmosphere.

After finally breaking off the game, you turn to discussion about the American economic system. Uneven distribution, lack of trust, unfair advantage for the "haves" are now very much on your mind. After trying to correct these injustices in the game of *Monopoly*, you have a new understanding about the difficulty of alleviating injustice in our society. The task appears impossible, but through this game you have learned that it is much more bearable if you and your colleagues are cooperating rather than competing.

NOTES

Theme Five: Happiness

Happy is the man who finds wisdom,
   and the man who gets understanding,
for the gain of it is better than
   gain from silver
   and its profits better than gold.
—Proverbs 3:13,14

Happiness is knowing what you are doing next; happiness is planning. Of course, happiness is still celebrating the now, the moment, but for our purposes in the following chapter, we will deal with the concept of happiness as the wisdom of good planning.

When an individual or group sets objectives and devises strategies for accomplishing these objectives, the groundwork has been laid for the development of a happy environment. Planning is not guaranteed to produce happiness, but it does give one a head-start toward achieving this elusive quality.

By planning one makes progress in stabilizing the future and, thereby, takes some of the uncertainty out of tomorrow. This is a liberating experience, because a person or an organization can depend upon certain things happening at a certain time. When one is freed from some of the momentary worries of what he is going to do right now, more openness occurs—an openness that allows happiness to enter one life or the group more easily.

THE NAME OF THE GAME:

Mapping

Order from ———————— J. J. Mar-Tam and Associates
1053 Delamont Ave.
Schenectady, New York 12307

Cost ———————— $25.00

Playing Time———————— Approximately 3 hours

Number of Participants ——— 15-25

PURPOSE:

MAPPING was developed primarily as a means of illustrating one particular planning model for a local church congregation. Underlying the purpose is the desire to communicate to local churches the importance of planning.

The planning model that is used is Strategic Planning, a model with the following major components: Assumptions, Objectives, Strategies, and Tactics. The same planning model is used by a number of industrial firms and educational institutions. The emphasis in the game and in the planning model is on participatory planning and planning on the basis of values.

A church that works seriously with this model will see the necessity for being able to articulate specific concrete values that are important and will also be committed to some intentional work in implementing these values. A church that attempts to implement Strategic Planning should also be prepared to broaden the base of participation in the decision-making process. Although different people wil participate at different levels at different times, the broader the base of participation the better.

Another specific objective of the game is to stress the importance of lay involvement in planning and to introduce a concept of lay ministry. Lay ministry in MAPPING refers to equipping laymen to witness to the Christian faith in their everyday lives. It means coming to terms with the institutional morality of the establishments they work for, the organizations they belong to, and the groups they are influenced by.

MAPPING represents a concept—a planning concept that is designed to help local congregations to better express their Christian faith.

COMPONENTS:

The most obvious functional needs of a church are the components of MAPPING. They are: Worship and Community Life, Christian Education, Corporate Ministry, Lay Ministry, and Stewardship of Human and Physical Resources.

Worship and community life is concerned with congregational involvement in preparation and implementation of worship and small group activities aimed at building community within the church. Experimental worship and the meaning of community are some of the concerns of the worship and community life task force.

The internal educational program of the church is the focus of the Christian education task force. Its primary concerns are the content, structure, and cost of the educational program.

Corporate ministry refers to the church as a total body moving out into the community. The corporate ministry task force is involved in deciding how much money will be given to foreign and local missions and what the nature of these missions will be.

The lay ministry task force deals with the relatively new concept of laymen ministering to the institutions that they are a part of during the week. The role of the church becomes that of equipping the laymen to witness to the Christian faith in their everyday lives.

The final task force is concerned with the job of keeping the facilities in working order and evaluating the stewardship of the physical resources of the church.

In each task force the emphasis is on deciding how much time (the clergyman's) and money is to be spent on a series of projects that represent policy decisions. Since there is a limited amount of money and time the task forces have to put their projects in priority and ultimately convince the total congregation of the value of the projects that they want to have funded.

THE PLAYERS:

Twenty-five persons is the optimum number for playing the game. The players are divided equally into the five task forces that represent the five major components listed above. Five team members create a good working unit.

The total group of players form the total congregation and they come together for congregational business meetings each quarter of the church year to ratify the work of the task forces.

The game can be easily administered by two people. The director of the game can serve as the church moderator and run the business meetings of the church as well as give general direction to the running of the gaming session. The second person will provide back-up help to the director. This person can be recruited by the director the evening the game is played and oriented in ten minutes. The director should allow at least three hours study time before running the game in order to become familiar with the various parts as well as the general nature of the game.

ENVIRONMENT:

MAPPING simulates a calendar year in the life of a church. The

138

game proceeds in rounds, with each quarter of the year representing a round. Each round includes a meeting of the task forces and also a congregational meeting. During the first round the congregation adopts a set of criteria by which to evaluate the Statement of Purposes that will be proposed in the succeeding round. In the second round the congregation will adopt a statement of purpose from three alternatives that are proposed. The individual task forces select the programs that are their individual priorities in the third quarter, and the final quarter of the church year is devoted to adopting the church's budget for the coming year. This process for adopting the budget is a participatory planning process that includes all the participants in a negotiating session from which the church's priorities will emerge.

The action for the game can become bogged down if the director does not keep it alive. There is a lot of reading required by the participants to stay abreast of the decision-making process. Although these represent some obstacles to be aware of, they are not insurmountable.

The most difficulty a local church will have is trying to use the planning model in its own congregation. My suggestion is that the church should spend a good deal of time following the game trying to apply the model to its local setting. Additional resource material on Strategic Planning can be obtained from the publisher of the game.

LEADER'S GUIDE

MAPPING is a simulation based on a model of Strategic Planning that was developed by Metropolitan Associates of Philadelphia. If you are not familiar with Strategic Planning, you should first read the enclosed book entitled "Strategic Planning for Church Organizations" by Richard Broholm.

MAPPING seeks to demonstrate how the theory that is worked out in that book can be operationalized. Since it is a game setting, some of the aspects of planning have to be over-simplified. However, we feel that the basic principles of planning are present in the game.

In this particular game the emphasis is put on the most tangible piece of any church's planning: namely, the budget. When a church actually does strategic planning it should spend a good portion of time developing assumptions. Appendix I seeks to deal with this important step which is passed over lightly in the game.

Other aspects of the game are stereotyped responses that are based on research compiled from "average" churches. Paradoxically, the church described in this game should not resemble any church now in existence and it should resemble thousands of churches now in existence. The Mapptown Community Church faces decisions that have been programmed for the purpose of illustrating Strategic Planning. However, we believe that the decisions are not totally unlike the decisions facing thousands of churches in all denominations.

The process of Strategic Planning and the rules of the game may seem a little confusing at first; however, if you read the material in the participant's guide and the game material you should have no trouble administering this game. We believe that it can be fun and even exciting.

The game is designed to be played with 15 to 25 persons and can be played in three and a half to four hours.

The four sequences outlined below represent the four quarters of the church year. Each sequence is divided into two parts: (a) the strategy team meetings and (b) the congregational quarterly business meetings.

Introducing the game:

It is impossible to adequately introduce any simulation game, and this one is no exception. Therefore, do not be concerned if people begin the game with some confusion about what they are to do; probably the less time spent in trying to introduce the game the better.

Divide the participants equally into the five strategy teams and ask each person to read the participant's guide. Answer any pressing questions. Again, try not to spend more than five or ten minutes answering questions.

SEQUENCE I. DEVELOPING CRITERIA.

After a brief word about the three areas of assumptions, explain the importance of a church's developing criteria by which to evaluate its proposed direction. In this model we generate those criteria from the assumptions. In order to make the game manageable we are using only a limited number of criteria. However, there is room for the teams to add and also to delete criteria they do not wish to use. For additional information on criteria see Appendix I.

Time Frame
Process Steps

10 Minutes A. Ask the participants to acquaint themselves with the criteria. They should ask, (1) are there criteria that should be added because of the special interest of our strategy team? (2) are there any criteria that should be deleted because we do not believe they should be included? If any of the groups have new criteria that they wish to add or a proposal to delete a particular criterion, they should be prepared to make a specific proposal to the congregation. (Important: The teams should not change the wording of the existing criteria. This is too time consuming.)

20 Minutes B. *Congregational meeting.* The process leader should act as Minister/Lay Leader and conduct the meeting. Merely accept recommendations regarding additions or deletions in the criteria. Staying within the twenty minutes, accept a motion for adoption of the criteria as they have been agreed upon. It may be helpful to use newsprint to record new criteria proposals.

SEQUENCE II. ADOPTING A STATEMENT OF PURPOSE.

15 Minutes A. Back in the strategy teams, each participant is given a copy of the three Statements of Purpose with their accompanying objectives. After reading through the three Statements of Purpose, each strategy team is asked to rate the statements on Form A, using the following scale: (Remember to add any new criteria that were adopted in Sequence I.)

> 0 - no correlation with the Statement of Purpose
> 1 - little correlation
> 3 - some correlation
> 5 - high correlation

If there is time following this procedure, the team may discuss which Statement of Purpose it wishes to recommend for adoption.

15 Minutes   B. *Congregational meeting.* The leader accepts motions as to which Statement of Purpose to adopt. Following discussion a vote is taken with each participant casting one vote. (It is not necessary, of course, for any participant to vote as his team votes or for him to support the Statement of Purpose which came out highest on the team's Form A.)

SEQUENCE III. ADOPTING AN ASKING BUDGET.

15 Minutes   A. Distribute to each team copies of the strategies that correspond with the Statement of Purpose that was adopted. Instruct each strategy team to select no more than three strategies to propose for the following year's budget. The strategies that are chosen by each team are recorded on Form B, with any explanatory notes.

15 Minutes   B. *Congregational meeting.* After each strategy team presents its proposed budget, the congregation votes on the adoption of the asking budget. (This vote should be perfunctory; if serious questions arise they can be dealt with during the negotiation.)

SEQUENCE IV. ADOPTING A BUDGET.

This sequence will be divided into at least two (maybe three) rounds, and each round will be divided into two parts. During the first part of the round the teams will choose the items they feel should go into the budget. The limit of the budget is $25,000; therefore, no team may exceed that amount in the total of the strategies that it chooses. For this first period in each round, the teams may communicate among one another by written messages only. For example, if one team felt it could gain support for one of its strategies by agreeing to support the strategy of another team, it would communicate that in the form of a written note to the team involved. The leader and the tabulator will serve as messengers for the teams during this period.

During the second period of each round the teams may communicate by having conferences. At least two rooms adjacent to the playing room should be set aside as conference rooms. A conference is requested on the form that the leader provides to those who request it. When a team fills out the form and the team or teams with whom a conference has been requested agree, the leader announces that a conference between Teams A and B will be held in Conference Room A. No more than two members of any team should attend a particular conference. The members of the team not in conference should continue the work of the group. When both conference rooms are filled, the leader waits until one is vacant before granting another conference. During the conference portion of the round, the teams may continue to communicate through written notes. They should not, however, talk among teams.

142

While the conference period is going on, the tabulator is calculating the results of the first round. At the beginning of the next round the teams are told on Form D which items in the budget have been passed.

Following is a suggested time sequence:

1st round: Decision forms due after 20 minutes
               Conference period - 10 minutes

2nd round: Decision forms due after 15 minutes
               Conference period - 10 minutes

3rd round: Decision forms due after 15 minutes
               Conference period - 10 minutes

If after three rounds of negotiating, the teams have not reached agreement on a budget, call a congregational meeting and ask for a motion that would settle the dispute. When such a motion is proposed, allow some time for debate and then take a vote. The budget should finally be passed by a majority vote of the congregation.

The time frame may appear to be too fast. However, our experience has shown that the game moves better and faster when you keep the pressure on to complete each task within the allotted time. The use of time in this game is another attempt to simulate reality. We all face time deadlines and finally decisions have to be made.

TABULATOR'S GUIDE

The person who serves as tabulator need not be familiar with the game prior to the run he participates in. It would not be impossible to tap someone at the beginning of the game and ask him to assist the leader.

During the first two sequences, he can observe and read the instructions outlined below and ask the leader questions for clarification.

At the end of the first half of the third sequence, the tabulator would take Form B from each of the five groups and put the strategies chosen by each team in the holder marked Asking Budget. Use the strategy inserts that are supplied for that purpose. Each team is then given one of the Asking Budgets at the conclusion of the Congregational Meeting in Sequence III.

During Sequence IV, the tabulator records the decisions of each of the Strategy Teams, which are reported on Form D. If in the first round of Sequence IV any strategy is supported by all five teams, it is considered "passed," and is recorded on Form E.

While the results of the first round are being tabulated, the teams will continue their negotiation through conferences. The tabulator should be able to report which strategies have passed within ten minutes after receiving the Decision Forms (Form D) from the five teams.

The same procedure will be followed during the second round of negotiation. During the first half of the round, while the teams are making their decision, the tabulator could help the leader by delivering messages, or he could perhaps put out a church newsletter.

During the third round of negotiation, only four teams have to vote for a strategy for it to be passed. The report form is the same.

For the innovative person the tabulator's role can be shaped in a number of very interesting and provocative ways. One possibility already mentioned is that he could publish a church newsletter to liven up the game and also to report pertinent information to the teams. Another possibility would be for the tabulator to become an advocate for a particular position in the early rounds of the game.

Also, if audio-video equipment is used, he may have to run this. But above all, the tabulator should not be afraid to innovate. Of course, he should clear signals with the leader to see that he is not hindering the process; however, creativity is very much a possibility in this role.

PARTICIPANT'S GUIDE

Welcome to MAPTOWN! For the next two or three hours you will be a layman/laywoman in the Community Church of Maptown. When the simulation game begins you will be assigned to a team, which will represent one of the five major strategy areas of your church.

If you have played other simulation games, you will know that games can be exciting as well as educational. Although this simulation game will not solve all your church's planning problems immediately, we do believe that it has value in communicating the importance of planning. More specifically, the objectives of this game are:

1. To communicate the importance of planning in the local church;
2. To illustrate how one planning model (Strategic Planning) operates; and
3. To make a strong pitch for lay involvement in planning.

The church for which you will be planning during the next year (four quarters will be simulated) has 275 members and a budget of $22,500. The church has not grown much in the past couple of years, but neither has it declined in membership or giving. Many people see the church as a crossroads; it needs a shot in the arm or a renewed sense of purpose.

In order to gain more concrete direction the governing body of the church has embarked on a new participatory planning process, called Strategic Planning. The date is early January and the church strategy teams are meeting for the first time. The first item on the agenda of each strategy team is consideration of the list of environmental, operational, and value assumptions that have been prepared by the Strategic Planning Committee. You will recognize many of these assumptions as ideas that were given by you and others in the questionnaire that the Strategic Planning Committee used to gather its information.

The game will progress in four sequences. Each sequence represents a quarter in the church year. The sequences will be divided into two parts: (a) strategy team meetings; and (b) congregational business meeting.

SEQUENCE I. DEVELOPING CRITERIA

During this sequence, the church will develop and adopt a set of criteria that will be used to evaluate the purpose of the church. The criteria will be the standard, the check list, which makes the church aware of what it wants to do. The criteria are developed from three categories of Assumptions: Environmental, Operational, and Theological/Value. See Appendix I for more information on this process.

SEQUENCE II. ADOPTING A STATEMENT OF PURPOSE

You will be given three alternative Statements of Purpose and, using the criteria adopted during the first quarter of the year, you will evaluate each purpose and finally adopt one as the direction the church will take for the coming year.

SEQUENCE III. ADOPTING THE ASKING BUDGET

After adopting a Statement of Purpose, each team will be given a list of strategies for implementing that particular statement. You will choose no more than three strategies from the ones you are given. These strategies will become part of the Asking Budget which will go to the entire church.

SEQUENCE IV. ADOPTING A BUDGET

Undoubtedly the Asking Budget will be more than the ten percent increase the church is anticipating in the giving next year, so during this sequence, the teams will negotiate among themselves for a budget totaling no more than $25,000.

A WORD OF CAUTION

Do not expect this simulation format to adhere to your preconceived notions about church planning. It is, we feel, a unique approach and worthy of your critical consideration. In order for you to get a feel for the process, we have posed clearly defined options for your church. Although the three Statements of Purpose that you will consider are not mutually exclusive (in fact, all three have elements of the other two), they are stated in a manner that clearly focuses on three different emphases for the church.

And the focusing is for a purpose; we believe that church planning should focus the resources of the church in an area where it is possible for the church to have an impact, rather than try to do everything in a limited way. You may become frustrated and puzzled at times, but you can also have fun experimenting with a new concept in church planning.

STATEMENT OF PURPOSE I.

The purpose of this church is to proclaim the love of Jesus Christ by calling men and women to accept him as their Lord and Savior and to bear witness to his power over all creation by supporting this church through active participation in worship, Sunday school, prayer meeting, missionary outreach, and personal evangelism.

OBJECTIVE I.

We will develop a program of Lay Ministry within the next year that will train laymen to witness to the saving love of God in the vocations and institutions in which they are involved.

OBJECTIVE II.

Through a traditional educational program we will provide training for children, youths and adults in the relevance of the historical biblical faith.

OBJECTIVE III.

We will continue to support the missionary outreach of this church by contributing of our resources to those who need to hear of the love of God.

OBJECTIVE IV.

We will integrate personal needs with the corporate experience of worship and fellowship so that members and friends of this church may discover a supporting fellowship of love and concern.

OBJECTIVE V.

Being aware that our human and material resources are a trust from God, we will use our material wealth—money, buildings, and time—to serve his purpose.

STATEMENT OF PURPOSE II.

The purpose of this church is to witness fearlessly to all persons that Jesus Christ as Lord and Savior is in the midst of human life calling us to decision and to make responsible use of our human and material resources. Our purpose for gathering for worship, study and planning is to celebrate God's presence in the world, to provide a community of support for one another, and to plan for mission.

OBJECTIVE I.

We will develop within the next year a program of Lay Ministry that will train laymen to witness to the redeeming and reconciling love of God in the vocations and institutions in which they are involved.

OBJECTIVE II.

Through a structured educational program we will provide training for children, youths and adults in the relevance of the historical biblical faith and how it equips us to act as Christians in our everyday involvements.

OBJECTIVE III.

We will continue to develop the mission action program of the church by contributing our resources to those persons and institutions that have pressing needs.

OBJECTIVE IV.

Through thoughtful study and experimentation we will strive to more effectively integrate the issues and concerns of our daily work with the way we celebrate on Sunday morning and as a community, in order that our work may truly be an extension and expression of our worship, and our worship the offering up of our work.

OBJECTIVE V.

Being aware that our human and material resources are a trust from God, we will use our material wealth—money, buildings, and time—to meet human needs.

STATEMENT OF PURPOSE III.

The purpose of this church is to serve as an agent of reconciliation between Jesus Christ and the world. The congregation, as a corporate body, shall move into the world as an agent of social change working in the world to bring God's Kingdom on earth in the form of a just and peaceful community.

OBJECTIVE I.

We will develop within the next year, a program of Lay Ministry that will train the laymen of the church to witness to the redeeming and reconciling love of God by moving into the world as a community of faith.

OBJECTIVE II.

Through a structured educational program we will provide training for children, youths and adults in the relevance of the historical biblical faith and how it calls God's community of faith to act in modern times.

OBJECTIVE III.

We will continue to develop the mission action program of the church by contributing our resources to those who are working for a more just society.

OBJECTIVE IV.

We will integrate personal needs and commitments with corporate experiences of worship and fellowship so that members and friends of this church may discover a dynamic fellowship of love and concern which sustains its actions in the world by celebrating God's radical demand upon this community of faith.

OBJECTIVE V.

Being aware that our human and material resources are a trust from God, we will use our material wealth—money, buildings, and time—to meet the needs of society.

STRATEGY TEAMS

I. LAY MINISTRY

The Lay Ministry strategy team is responsible for equipping lay men and women in the church for the task of acting on their Christian principles in their secular lives.

II. CHRISTIAN EDUCATION

Providing an education program for the church in which children, youths and adults can learn the historic roots of the Christian faith and systematically reflect on its present day application is the primary task of the Christian Education strategy team.

III. CORPORATE MINISTRY

The Corporate Ministry strategy team is to develop the church's agenda for united action around the most pressing needs of the world in which, as a corporate body, we can responsibly and effectively become involved.

IV. WORSHIP AND COMMUNITY LIFE

To plan and provide for worship experiences for the church is one of the primary tasks of this strategy team. Also the Worship and Community Life strategy team is responsible for initiating programs that undergird the development of a community of faith.

V. STEWARDSHIP OF HUMAN AND MATERIAL RESOURCES

The Stewardship of Human and Material Resources strategy team is responsible for seeing that the essential services of the church are administered, that the building is maintained, and that all resources are used as a trust from God to meet basic human needs.

CRITERIA

Operational Criteria:

1. Does it support our denomination in those areas where the denominational work is consistent with the church's mission?

2. Does it serve to broaden the fellowship of Christian concern by uniting more men and women in a community of faith?

3. Does it allow for all men and women to participate to the fullest extent of their interest and ability in the policy decisions of the church?

4. Does it reflect an engagement in mission in such a way that members and friends of the church will want to contribute financially?

5. Does it provide for better and more efficient use of our physical facilities?

6. Does it call for a celebrating of God's act of redeeming mankind?

Environmental Criteria:

1. Does it counteract the exhaustion of natural resources and the increased pollution of the environment?

2. Does it help facilitate a more equal redistribution of the goods and services basic to human welfare?

3. Does it communicate that all men are united under God despite cultural differences, such as youth culture and prevailing culture?

4. Does it take cognizance of the fact that nations are under the judgment of God and that natural unity is dependent upon a just social order?

5. Does it recognize that God is present in the secular structures of society and that man's work is as sacred as his worship?

6. Does it equip men and women to pursue their rights in shaping their own destiny?

THEOLOGICAL/VALUE ASSUMPTIONS

1. God has taken the initiative in creating the world and man is called to respond as a *steward of the earth and its fruits.*

2. God through Jesus Christ took the initiative in redeeming mankind, and man is called to respond in *thankful acceptance and celebration.*

3. God, through the Holy Spirit, took the initiative in uniting all men and man is called to respond by *relating to others in love.*

4. God has provided for man's well-being and man is called to respond by *sharing his goods and services for the welfare of others.*

5. God judges the nations and man is called to respond by *working for justice through social and institutional change.*

6. God took the initiative in liberating men and man is called to respond by *exercising his right to decide his own future.*

7. God has taken the initiative in dignifying man's labor, and man is called to respond by *fulfilling his calling through work.*

8. God has founded the church and man is called to respond by *engaging in her mission.*

OPERATIONAL ASSUMPTIONS

* x1. The church has denomination ties that it wishes to maintain. (#8)**

  x2. The policy decisions of the church are to be made by the total church community. There is a democratic form of government. (#6)

  x3. The financial support of the church comes exclusively from the church members and certain friends of the church. (#8)

  x4. The buildings of the church are used six to ten hours a week (other than the church office). (#6)

  5. The governing body of the church consists of representatives of a cross section of the congregation.

  6. The church budget will increase about 10% next year.

  x7. The number of members of the church is remaining static. (#3)

  8. The church facilities are now in good repair, but some major capital improvements will need to be made on the Sanctuary within the next five years.

  x9. Sunday morning worship is helpful to the members of the church. (#2)

       * Those assumptions that are marked "x" are the ones that have been converted to criteria.

       ** The numbers refer to the number of the theological/value assumption it has been intersected with in order to create a criteria.

ENVIRONMENTAL ASSUMPTIONS

* x1. The natural resources of man are being exhausted and the atmosphere is becoming more polluted. (#1)**

  2. Technological advancement is creating a more affluent society.

  x3. The generation gap is being felt more strongly. (#3)

  x4. The gap between the rich and the poor is not being bridged. (#4)

  x5. The American political arena remains polarized and potentially unstable. (#5)

  6. The nations of the world become even more intimately interdependent.

  7. Nuclear war threatens all civilization.

  8. The large specialized organization dominates the social field.

  9. The population continues to explode.

x10. America is becoming more city-like and secularized. (#7)

11. Efforts to expand the development of national economies are continuing.

12. Science and technology are reshaping society.

x13. Men and women feel less capable of participating in shaping their future. (#6)

> * Those assumptions that are marked "x" are those that have been converted to criteria.

> ** The numbers refer to number of the theological/value assumption it has been intersected with in order to create a criteria.

CONVERTING ASSUMPTIONS TO CRITERIA

There are three simple steps to converting the Environmental and Operational Assumptions into Criteria that are used to weigh and evaluate the Statement of Purpose, Objectives, and Strategies.

First, identify which Environmental trends and Operational assumptions are important and should serve as criteria.

Second, intersect them individually with an appropriate theological/value assumption.

Third, write the trend/assumption as a Criteria.

Following are four examples: (two from each category)

Operational Assumption
. . . The church has denominational ties that it wishes to maintain.

Value Assumption
. . . God has founded the church and man is called to respond by engaging in her mission.

Criteria
. . . Does it support our denomination in those areas where the denominational work is consistent with the church's mission?

Operational Assumption
. . . The number of members of the church is remaining static.

Value Assumption

. . . God, through the Holy Spirit, took the initiative in uniting all men and man is called to respond by relating to others in love.

Criteria

. . . Does it serve to broaden the fellowship of Christian concern by uniting men in a community of faith?

Environmental Assumption

. . . The natural resources of man are being exhausted and the atmosphere is becoming more polluted.

Value Assumption

. . . God has taken the initiative in creating the world and man is called to respond as a steward of the earth and its fruits.

Criteria

. . . Does it serve to counteract the exhaustion of natural resources and the increased pollution of the atmosphere?

Environmental Assumption

. . . The gap between the rich and the poor is not being bridged.

Value Assumption

. . . God has provided for man's well-being and man is called to respond by sharing his goods and services for the welfare of others.

Criteria

. . . Does it help facilitate a more equal redistribution of goods and services basic to human welfare?

STATEMENT OF PURPOSE I
OBJECTIVE I.

We will develop a program of Lay Ministry within the next year that will train laymen to witness to the saving love of God in the vocations and institutions in which they are involved.

		COST	MINISTER'S TIME (hrs/wk)
STRATEGY 1.	To run four classes (six weeks each) in "how to win others to Christ."	$1,500.00	7
Tactics:	a. The minister will recruit the first class of twelve. b. The second class will be recruited by the first class, etc. c. The minister shall teach all the classes.		
STRATEGY 2.	To invite twenty-five selected young men in groups of five to consider the ministry.	500.00	2
Tactics:	a. The minister and the chairman of the board shall select these young men. b. They and their parents will be invited to participate in a series of meetings designed to encourage them to enter full-time Christian service.		
STRATEGY 3.	To train ten lay people to effect humanizing change in their place of work.	500.00	None
Tactics:	a. The chairman of the lay strategy committee selects ten people to begin studying how the Christian faith affects their work. b. The religion professor of the local university is		

invited to lead the class.

c. The participants are asked to adopt an action goal at the end of six sessions.

STRATEGY 4. To establish five different sectors in the congregation to begin the process of interpreting lay ministry.   1,000.00   4

Tactics: a. Divide the congregation into five geographical sections.

b. Set up home meetings in each of the sectors.

c. Prepare a multi-media presentation on the rationale for and potential of lay ministry.

STATEMENT OF PURPOSE I
OBJECTIVE II.

Through a traditional educational program we will provide training for children, youths and adults in the relevance of the historical biblical faith.

		COST	MINISTER'S TIME (hrs/wk)
STRATEGY 1.	Run a one hour Sunday school for children, youths and adults each week.	$3,000.00	None
Tactics:	a. The Sunday school superintendent shall recruit the teachers and order the materials from the denominational press.		
	b. The Sunday school secretary shall keep track of attendance and send cards to those who have missed more than three consecutive weeks.		
	c. The Sunday school treasurer shall keep track of the money and disburse it as the church directs.		
STRATEGY 2.	Run a youth program for the Junior Highs and Senior Highs each Sunday evening.	2,000.00	7
Tactics:	a. The pastor shall recruit adult advisers for the youth groups and meet with the groups.		
	b. The advisers and the youth leaders shall work out the content of the program.		
STRATEGY 3.	Begin a young couples class.	500.00	None
Tactics:	a. Pastor shall have a social for the young cou-		

ples in the church.

    b. A respected church leader shall be recruited to teach this new Sunday school class.

STRATEGY 4. Cooperate with three churches in the area for the establishment of a unified weekday educational program.   1,500.00   None

Tactics: a. Establish a committee to explore the idea with other churches.

    b. Begin a pilot project for Junior Highs two days a week after school.

STATEMENT OF PURPOSE I
OBJECTIVE III.

We will continue to support the missionary outreach of this church by contributing our resources to those who need to hear of the love of God.

	COST	MINISTER'S TIME (hrs/wk)
STRATEGY 1. Contribute 22% of the budget to the denomination's budget for the purpose of supporting missionaries in the field.	$5,500.00	None

Tactics:
a. Instruct the church treasurer to pay the 22% on a quarterly basis.
b. Appoint a special committee to examine carefully how the denomination spends the money it is given.

STRATEGY 2. Contribute to the local chapter of Campus Crusade for Christ.	500.00	None

Tactics:
a. The minister shall appoint a committee to bring an evangelistic team to the local college campus.
b. Begin identifying prospects for the Crusade.

STRATEGY 3. Contribute to a special project to send a missionary to the inner city to set up a storefront church to preach the "Good News" of the Gospel.	500.00	None

Tactics:
a. Minister shall contact the people working on this to express the church's interest.
b. The ladies missionary society shall follow-up and recommend how

much should be given to this project.

c. Anything more than the budget amount shall be given by the missionary society.

STRATEGY 4. Establish a low income housing project on the vacant land next to the church.   1,500.00   None

Tactics: a. Establish a nonprofit corporation to accomplish this.

b. Work with the local housing authority to obtain rent supplement for the project.

c. Use the church to extend a line of credit to the nonprofit housing corporation.

STATEMENT OF PURPOSE I
OBJECTIVE IV.

We will integrate personal needs with the corporate experience of worship and fellowship so that members and friends of this church may discover a supporting fellowship of love and concern.

		COST	MINISTER'S TIME (hrs/wk)
STRATEGY 1.	To provide a challenging worship experience on Sunday mornings.	$4,500.00	15
Tactics:	a. The pastor shall prepare for and lead this service. b. A part-time choir director and organist shall see that special music is provided for the morning worship.		
STRATEGY 2.	The pastor, through home visits and personal counseling, shall provide support and relief to those in personal need.	2,500.00	12
Tactics:	a. The pastor shall call on the shut-ins at least once a month. b. The pastor shall ask the Board of Deacons to help with home visits. c. The pastor shall make himself available to members of the church for at least one hour a day for counseling.		
STRATEGY 3.	To provide home-centered Bible study.	1,000.00	4
Tactics:	a. The church shall print an outline for daily Bible reading and devotion for each member of the congregation. b. Pastor shall provide Bible study in one home		

161

each week for members of congregation who live in that area.

STRATEGY 4. Set up five small groups (led by minister or some person with training) to experiment with developing communities of support.   500.00   None

Tactics: a. The minister shall organize two groups during the first quarter of the year.
b. Two more groups shall be organized after that.
c. The composition of the groups shall vary: for example, one group of young couples, one group of women, one group of men, one group chosen at random, one group based on geography.

STATEMENT OF PURPOSE I
OBJECTIVE V.

Being aware that our human and material resources are a trust from God, we will use our material wealth—money, buildings, and time—to serve his purpose.

		COST	MINISTER'S TIME (hrs/wk)
STRATEGY 1.	To maintain and administer essential services, property, and equipment.	$5,000.00	12
Tactics:	a. Hire a custodian. b. Buy cleaning supplies. c. Maintain equipment (typewriters, pianos, etc.). d. Pastor shall administer church program and resources.		
STRATEGY 2.	Undertake a program of capital improvement.	2,500.00	None
Tactics:	a. Expand the parking lot. b. Install a carpet in the parsonage.		
STRATEGY 3.	Establish a capital reserve fund in case the church falls on hard times in the future.	1,500.00	None
Tactics:	a. Put 5% of present budget in this fund to initiate it. b. Encourage people to leave money in their wills to this fund.		
STRATEGY 4.	Draw up a plan for more efficient use of the church facilities.	1,000.00	None
Tactics:	a. Devise plans for different kinds of uses for the buildings, such as a school or a community center. b. Employ an architect to lay out time and cost of some alternatives.		

STATEMENT OF PURPOSE II
OBJECTIVE I.

We will develop within the next year a program of Lay Ministry that will train laymen to witness to the redeeming and reconciling love of God in the vocations and institutions in which they are involved.

		COST	MINISTER'S TIME (hrs/wk)
STRATEGY 1.	To train 20 lay people to effect humanizing change within their place of work.	$3,500.00	15

Tactics:
a. The minister shall begin a 25 hour training program for those who want to effect change within their place of work.
b. After the 25 hour training, each member shall develop an action plan and begin to work for institutional change in his place of work.
c. The group of 20 shall continue to meet and give support to each other.

STRATEGY 2. To set aside $2,000 to provide moral and financial help to those who have put their education/jobs on the line due to efforts in lay ministry.   2,000.00

Tactics:
a. The Lay Ministry committee/board shall establish a three-man team to administer these funds.
b. Disbursements are subject to the approval of the entire Lay Ministry group.

STRATEGY 3. To establish five different sectors in the congregation to begin the process of interpreting Lay Ministry.   1,000.00   4

Tactics: a. Divide the congregation into five geographical sections.
b. Set up home meetings in each of the sectors.
c. Prepare multi-media presentation on the rationale for and potential of Lay Ministry.

STRATEGY 4.   To begin a congregation-wide education program regarding the problems of war, race and pollution.   1,000.00   2

Tactics: a. Divide the congregation into five geographic sectors and have member of Lay Ministry group organize each sector around an issue.
b. Establish a Sunday morning forum on issues.

STATEMENT OF PURPOSE II
OBJECTIVE II.

Through a structured educational program we will provide training for children, youths and adults in the relevance of the historical biblical faith and how it equips us to act as Christians in our everyday involvements.

		COST	MINISTER'S TIME (hrs/wk)
STRATEGY 1.	Run a one-hour Sunday school program for children, youths and adults each week.	$3,000.00	None
Tactics:	a. The Sunday school superintendent shall recruit the teachers for the various classes.		
	b. Each teacher, after consultation with her class, and the minister shall order materials to be used during that particular quarter.		
	c. All money collected in the Sunday school classes will be used for the operations budget of the church.		
STRATEGY 2.	Set up an experimental youth program.	2,000.00	None
Tactics:	a. Hire a special youth minister to relate to the kids outside as well as inside the church.		
	b. Provide facilities during the week for youths to use.		
	c. Provide program money for the part-time youth minister.		
STRATEGY 3.	Cooperate with three churches in the area for the establishment of a unified weekday educational program.	1,500.00	None

Tactics: a. Establish a committee to explore the idea with other churches.
b. Begin a pilot project for Junior Highs two days a week after school.

STRATEGY 4. Cooperate with other churches in the area to provide a free school for people interested in becoming more informed on the issues of the day.   1,000.00   None

Tactics: a. Set up a committee to explore the idea with other churches.
b. Begin a pilot program by offering two courses each quarter for this next year in such things as: Black History, History of U.S. Imperialism, and Survival: Will General Motors Do Us In?

STATEMENT OF PURPOSE II
OBJECTIVE III.

We will continue to develop the mission action program of the church by contributing our resources to those persons and institutions that have pressing needs.

		COST	MINISTER'S TIME (hrs/wk)
STRATEGY 1.	Contribute 22% of the church budget to the denomination's unified budget.	$5,500.00	None
Tactics:	a. Instruct the treasurer to pay the 22% on a monthly basis.		
	b. Invite a spokesman for the denomination to come to the church to discuss with the congregation how the money is being spent.		
STRATEGY 2.	Establish a low income housing project on the vacant land next to the church.	1,500.00	None
Tactics:	a. Establish a non-profit corporation to accomplish this.		
	b. Work with the local housing authority to obtain rent supplement for the project.		
	c. Use the church to extend a line of credit to the nonprofit housing corporation.		
STRATEGY 3.	Contribute to institutions that are working to relieve the suffering of those in need.	500.00	None
Tactics:	a. Contribute to the American Friends Service Committee.		
	b. Contribute to the neigh-		

168

borhood community center.

STRATEGY 4. Contribute to the local Council of Churches "Agenda for Action" program.   1,000.00   None

Tactics: a. Appoint a member of that Council of Churches committee.
b. Try to involve the congregation in the implementation of the program.

STATEMENT OF PURPOSE II
OBJECTIVE IV.

Through thoughtful study and experimentation we will strive to more effectively integrate the issues and concerns of our daily work with the way we celebrate on Sunday morning and as a community, in order that our work may truly be an extension of our worship, and our worship the offering up of our work.

		COST	MINISTER'S TIME (hrs/wk)
STRATEGY 1.	To provide an inspiring, challenging worship experience on Sunday mornings.	$4,500.00	15
Tactics:	a. The minister shall work with the worship and community life group to plan and lead a service that meets the objective.		
	b. Music shall be under the supervision of a special committee, but emphasis shall be placed on total congregational involvement.		
STRATEGY 2.	Through personal counseling the minister and the church shall provide support and relief to those in special need.	1,000.00	4
Tactics:	a. The deacons shall see that the shut-ins are visited once a month.		
	b. The minister shall be available three hours each week for personal counseling.		
STRATEGY 3.	Set up five small groups (led by minister or some person with training) to experiment with developing communities of support.	5,000.00	None

Tactics: a. The minister shall organize two groups during the first quarter of the year.
  b. Two more groups shall be organized after that.
  c. The composition of the groups shall vary: For example, one group of young couples, one group of women, one group of men, one group chosen at random, one group based on geography.

STRATEGY 4. Set up small groups to experiment with different types of community lifestyles.   1,500.00   4

Tactics: a. Organize a Christian commune.
  b. Release minister part-time to organize a youth hostel in the community.
  c. Organize a group of young couples to do cooperative buying of goods.

STATEMENT OF PURPOSE II
OBJECTIVE V.

Being aware that our human and material resources are a trust from God, we will use our material wealth—money, buildings, and time—to meet human needs.

		COST	MINISTER'S TIME (hrs/wk)
STRATEGY 1.	To maintain and administer essential services, property, and equipment.	$5,000.00	12
Tactics:	a. Hire a custodian. b. Buy cleaning supplies. c. Maintain equipment.		
STRATEGY 2.	Begin paying taxes to the township for fire and police protection and other city services.	1,000.00	
Tactics:	a. Negotiate with the city or township for a fair amount to pay. b. Stewardship committee shall inform other churches in the area of this action.		
STRATEGY 3.	Draw up plans for sale of the church property.	1,500.00	
Tactics:	a. Have the building appraised. b. Devise alternative uses for the building or the land, such as tearing it down and building an apartment house (with one portion used by the church), or converting the building to a school.		
STRATEGY 4.	Establish a capital reserve fund in case the church falls on hard times in the future.	1,500.00	
Tactics:	a. Put 5% of present budget in this fund to initiate it. b. Encourage people to leave money in their wills to this fund.		

STATEMENT OF PURPOSE III
OBJECTIVE I.

We will develop within this next year, a program of Lay Ministry that will train the laymen of the church to witness to the redeeming and reconciling love of God by moving into the world as a community of faith.

		COST	MINISTER'S TIME (hrs/wk)
STRATEGY 1.	To initiate a "peace-making task force" for the purpose of informing the congregation of the action it should take with regard to the war.	$ 500.00	None
Tactics:	a. Ask the local chairman of SANE to head this task force. b. Take responsibility for distribution of literature in the church.		
STRATEGY 2.	To begin a congregation-wide education program regarding the problems of war, race and pollution.	1,000.00	None
Tactics:	a. Divide the congregation into five geographic sectors and have members of Lay Ministry group organize each sector around an issue. b. Establish a Sunday morning forum on issues.		
STRATEGY 3.	Release minister to spend ¼ time working with college students trying to establish a draft counseling center.	3,500.00	15
Tactics:	a. Provide support for the minister by working to recruit volunteers for the project. b. Pledge $500 (in addition to church giving)		

for program money for the project.

STRATEGY 4. To train 10 lay people to effect humanizing change in their place of work.   500.00   None

Tactics: a. The chairman of the lay strategy committee selects 10 people to begin studying how the Christian faith affects their work.
b. The religion professor of the local university is invited to lead the class.
c. The participants are asked to adopt an action goal at the end of six sessions.

STATEMENT OF PURPOSE III
OBJECTIVE II.

Through a structured educational program we will provide training for children, youths and adults in the relevance of the historical biblical faith and how it calls God's community of faith to act in modern times.

	COST	MINISTER'S TIME (hrs/wk)
STRATEGY 1. Run a one-hour Sunday school program for children, youths and adults each week.	$3,000.00	None

Tactics:
a. The superintendent shall recruit the teachers for the various classes.
b. Each teacher, after consultation with her class and the minister, shall devise an action project for the class.
c. All money collected through the Sunday school shall be given to the Council of Churches "Agenda for Action" program.

STRATEGY 2. Organize a group of high school and college youth to set up work camps and other projects related to the major issues of the day.	3,500.00	15

Tactics:
a. Release the minister ¼ of the time to work for the local peace organization to accomplish this task.
b. Provide financial and moral support to the group.

STRATEGY 3. Cooperate with other churches in the area to provide a free school for

	people interested in becoming more informed on the issues of the day.	1,000.00	
Tactics:	a. Set up a committee to explore the idea with other churches.		
	b. Begin a pilot program by offering two courses each quarter for this next year in such things as: Black History, History of U.S. Imperialism, and Survival: Will General Motors Do Us In?		

STRATEGY 4.	Begin a young couples class.	500.00	None
Tactics:	a. Pastor shall have a social for the young couples in the church.		
	b. A respected church leader shall be recruited to teach this new Sunday school class.		

STATEMENT OF PURPOSE III
OBJECTIVE III.

We will continue to develop the mission action program of the church by contributing our resources to those who are working for a more just society.

		COST	MINISTER'S TIME (hrs/wk)
STRATEGY 1.	Contribute 22% to the denomination's unified budget.	$5,500.00	None
Tactics:	a. Instruct the treasurer to send the 22% on a bi-monthly basis to the denomination. b. Send a list of questions to the denominational headquarters to find out how the money is being spent.		
STRATEGY 2.	Contribute to the local Council of Churches "Agenda for Action" program.	1,000.00	None
Tactics:	a. Appoint a member of that Council of Churches committee. b. Try to involve the congregation in the implementation of the program.		
STRATEGY 3.	Contribute to action projects that are working for justice in the society.	500.00	None
Tactics:	a. Contribute to the local peace center. b. Contribute to the American Friends Service Committee. c. Contribute to the local anti-poverty agency.		
STRATEGY 4.	Contribute to a special project to send a mis-		

sionary to the inner city to set up a store front church to preach the "Good News" of the Gospel.   500.00   None

Tactics: a. Minister shall contact the people working on this to express the church's interest.

b. The ladies missionary society shall follow-up and recommend how much should be given to this project.

c. Anything more than the budgeted amount shall be given by the missionary society.

STATEMENT OF PURPOSE III
OBJECTIVE IV.

We will integrate personal needs and commitments with corporate experiences of worship and fellowship so that members and friends of this church may discover a dynamic fellowship of love and concern which sustains its actions in the world by celebrating God's radical demand upon this community of faith.

		COST	MINISTER'S TIME (hrs/wk)
STRATEGY 1.	To provide a dynamic and challenging worship experience on Sunday mornings.	$3,500.00	10
Tactics:	a. The minister shall work with the worship and community life group to plan and lead a service that is consistent with the objective.		
	b. Special music shall emerge from a variety of mediums as the need arises.		
STRATEGY 2.	Experiment with a combination of worship and action at least once a month.	1,000.00	4
Tactics:	a. Hold a worship at the local draft board on the first Sunday of the month until the draft is ended.		
	b. On two or three occasions join with a black congregation for Sunday morning worship.		
STRATEGY 3.	Set up small groups to experiment with different types of community lifestyles.	1,500.00	4
Tactics:	a. Organize a Christian commune.		
	b. Release minister part-time to organize a youth		

hostel in the community.

c. Organize a group of young couples to do co-operative buying of goods.

STRATEGY 4. Set up five small groups (led by minister or some person with training) to experiment with developing communities of support. 500.00

Tactics: a. The minister shall organize two groups during the first quarter of the year.

b. Two more groups shall be organized after that.

c. The composition of the groups shall vary: for example, one group of women, one group of men, one group chosen at random, one group based on geography.

STATEMENT OF PURPOSE III
OBJECTIVE V.

Being aware that our human and material resources are a trust from God, we will use our material wealth—money, buildings and time—to meet the needs of society.

		COST	MINISTER'S TIME (hrs/wk)
STRATEGY 1.	To maintain and administer essential services, property, and equipment.	$3,000.00	12 hours— administration
Tactics:	a. Assign each family (that could do it) in the church the job of cleaning for the period of one week.		
	b. Buy cleaning supplies.		
	c. Maintain equipment, such as typewriters, pianos, etc.		
	d. Minister shall administer the program and resources of the church.		
STRATEGY 2.	Undertake a program of capital improvement.	2,000.00	
Tactics:	a. Remodel the church house to be used as a youth center.		
	b. Buy cleaning supplies.		
STRATEGY 3.	Draw up a plan for more efficient use of the church facilities.	1,000.00	None
Tactics:	a. Devise plans for different kinds of uses for the buildings, such as a school or a community center.		
	b. Employ an architect to lay out time and cost of some alternatives.		
STRATEGY 4.	Begin paying taxes to the township for fire and po-		

lice protection and other city services.   1,000.00   None

Tactics: a. Negotiate with the city or township for a fair amount to pay.

b. Stewardship committee shall inform other churches in the area of this action.

COMMENTS:	III	II	I	STATEMENT OF PURPOSE	FORM A
				Denominational Times	
				Involve More People in the Church	
				Participation in Policy Making	
				Engagement in Mission / Financial involvement	
				Use of Facilities	
				Celebration	
				Natural Resources / Pollution	
				Distributing Goods To Meet Human Equalization of Wealth Need	
				Bridge Generation Gap	
				Help Achieve Social Justice	
				God at Work in Secular Structures	
				Equip Us To Shape Our Own Destiny	
				TOTAL	

ASKING BUDGET REPORT FORM

Strategy Team_____

Following are the strategies under our objectives that we would like to accomplish next year:

Strategy Number   Cost   Minister's Time

Comments: (Why should the other strategy teams consider these strategies seriously?)

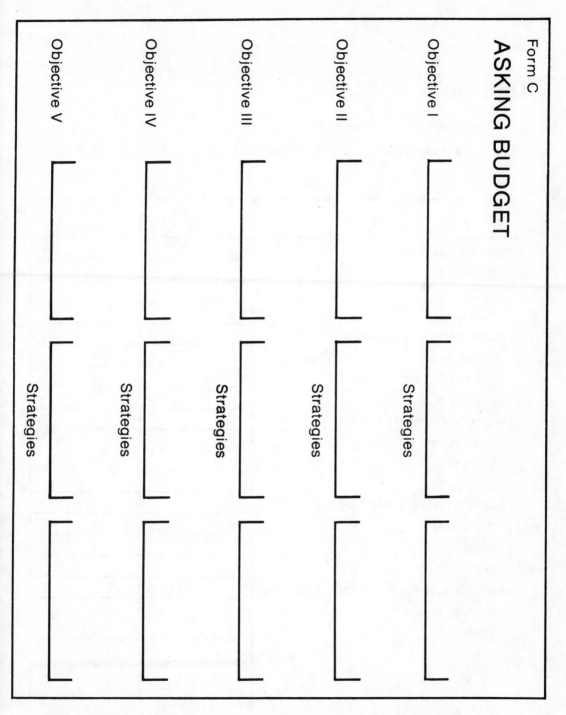

Form C

ASKING BUDGET

Objective I

Objective II

Objective III

Objective IV

Objective V

Strategies

Strategies

Strategies

Strategies

Strategies

Chart should be large enough so that a 3 x 5 card can be inserted in each pocket.

FORM D

DECISION FORM

Team_____   Round_____

Check the strategies that you wish to support. The total must not exceed $25,000.

	Cost	Minister's Time
Objective I. Lay Ministry		
____Strategy 1.		
____Strategy 2.		
____Strategy 3.		
Objective II. Christian Education		
____Strategy 1.		
____Strategy 2.		
____Strategy 3.		
Objective III. Corporate Ministry		
____Strategy 1.		
____Strategy 2.		
____Strategy 3.		
Objective IV. Worship and Community Life		
____Strategy 1.		
____Strategy 2.		
____Strategy 3.		
Objective V. Material and Human Resources		
____Strategy 1.		
____Strategy 2.		
____Strategy 3.		
TOTALS:		

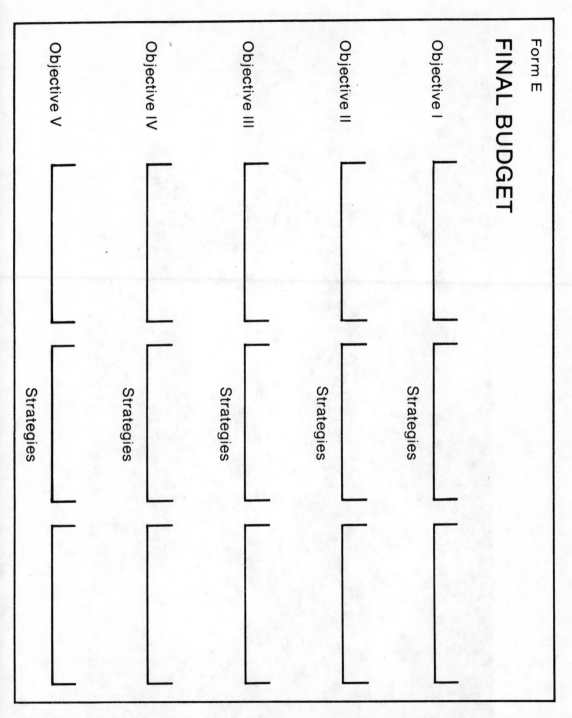

Form E

FINAL BUDGET

Objective I

Objective II

Objective III

Objective IV

Objective V

Strategies

Strategies

Strategies

Strategies

Strategies

Chart should be large enough so that a 3 x 5 card can be inserted in each pocket.

THE NAME OF THE GAME:

Plans

Order from —————————— Western Behavioral Science Institute
SIMILE II
P.O. Box 1023 - 1150 Silverado
La Jolla, California 92037

Cost ————————————— $3.00 sample kit
35.00 25 student kit
50.00 35 student kit

Playing Time——————— 4 - 6 hours

Number of Participants ——— 12-36

PURPOSE:

PLANS provides an opportunity for students to better understand the procedure of lobbying in the national Congress. For the purpose of this simulation six lobbying forces are singled out, their primary goals identified, and the means by which they seek to achieve their self-interest examined. Perhaps the single-most important aspect of this simulation is its treatment of the role of national policy. The way well-financed special interest groups affect the decisions Congress makes is often overlooked. Why it takes decades to decrease the oil depletion allowance, years to decrease the amount of pollution automobiles dump into the air, months to end crippling labor-management disputes, is the result of the power of special interest groups. PLANS is a game that seeks to help students understand this important and powerful role in national politics.

COMPONENTS:

The six lobbying forces used are: the military, the nationalists, business, the internationalists, labor, and civil rights groups. The nationalists and the military cover the American political right, while the civil rights groups and the internationalists represent the American political left.

Business and labor swing depending upon their own self-interest. More than any of the other groups, they are immediately affected by the dollars-and-cents decisions that are reached. One could argue that the civil rights groups have more riding on what happens than the other groups. However, like most civil rights proposals, those in this game that are favorable to blacks and minorities have more ceremonial than actual value. Therefore, the groups to watch in PLANS are business and labor.

Since so much of the action depends on business and labor, some of the reality is lost because the simulation cannot account for the divergence within the business and labor movements respectively. This weakness is rather small when considered in the context of the overall purpose of PLANS, which is not to study the lobbying forces individually, but to begin to understand the way in which they collectively influence national policy.

The national policy that the special interest groups influence ranges from the size of military spending to civil rights legislations, from the U.N. budget to minimum wage, and from federal aid to education to rate of income tax. The issues discussed (in the form of proposals) are broad—perhaps too broad—but they do represent a good cross section of problems that combine to formulate policy—national and international.

The decisions that are made on the proposals are reflected in a series of 18 national indicators. These include the G.N.P., profit before taxes, the ratio of non-white to white per capita income, etc. The value of these indicators is in learning their importance in real life. The indicators move in the game only when a proposal is passed and because of this do not figure significantly into the dynamics of the game. However, each indicator is quite important in real life and should be examined by the class in follow up of this simulation game.

THE PLAYERS:

The number of persons playing PLANS at one time can range from 12 to 36. Since the participants are divided equally into six teams (each representing one of the lobbying groups), this means that the team size can range from 2 to 6.

Although there is much debate about team size, I prefer 3 per team. I believe that this provides enough people for a good dynamic inter-team relationship, but is small enough for each person to participate fully. I also feel that it is better to have an odd number on a team. In other words, 5 persons per team is better than 4. There is a psychological plus about knowing that the decision-making process is not going to wind up deadlocked.

The concepts that are discussed in this game are more abstract than the concepts in a game like CRISIS or URBAN DYNAMICS. Therefore, it is better understood by juniors and seniors in high school. The proposals that the players must decide on are not the type of issues that they read about in headlines or in their current-events newspaper. In PLANS they must deal with the subtleties and the undercurrents of national policy.

The game operator may have to use some of the students to help run the game. One drawback to wide use of PLANS is that it takes 5 persons to administer it. A game operator directs the action and assists 1 or 2 others in delivering messages. It takes, however, 3 persons to do the calculating. The PLANS model is more complicated than necessary but it must be followed. Two people record the points from the teams and calculate their effects. A third calculator publishes a yearly news bulletin which is the way the results of the decisions are announced. Whoever is used to assist in calculating must spend about two hours prior to the game studying and practicing his/her responsibilities. The messengers can be recruited on the spot.

ENVIRONMENT:

The physical environment of PLANS is totally artificial because there is no real life counterpart to the setting that is created in the game. For CRISIS the U.N. is the real-life counterpart; for *Sitte* it is the city council meeting; but for PLANS the game setting is merely a room set up for six teams that have to bargain.

If you have a large room the teams can be placed far apart. Be sure to put teams that have the most in common farthest apart. The simulation creates its own blocks to direct communication (Communicate only through written messages and scheduled conferences.), but the game director must work hard to maintain the natural distance between the interest groups.

Although it will be necessary for the game director to prevent unofficial communication among the groups, he/she may have to prod some official communication. Work closely with each group and see that it understands its goals and that it knows who is both for and against it. The messages that circulate among the teams should contain some humor, but for the most part the participants should be encouraged to work on real solutions. If the players really get into the game, they will produce their own good humor within the context of the material on which they are working.

The content of the game material is set in 1965. The players are given a "Current State of Affairs and Plans Form" which indicates the national and international situation when the game begins. Although each round represents one year, the game proceeds in six month periods (or half periods) and periodic up-dates keep the participants informed about the effect of the decisions they are making.

Although the game is somewhat complicated (particularly the national indicators aspect of the game), I consider it the best of the Simile II games. The material that is used is solid, the action is real, and the subject treated is critical to understanding and changing national policy.

Sitte Model

The basic framework of the game *Sitte*, discussed in the theme of Freedom, can be used as a structure for actual decision-making and planning by those interest groups within a given organization who are attempting to reach consensus about a future course of action.

Actual proposals or plans, developed by each interest group meeting separately over a period of time, would be distributed to all in advance of a conference day. Some preliminary negotiation would probably go on in advance as well. On the given day, everyone would gather for several rounds of negotiations interspersed with vote-casting for one or more of the contending proposals.

After a limited period of time those proposals receiving greatest coalition support would be recognized as priority proposals.

The results of the "play period" then could serve as data for a final decision-making meeting at which the "players" would reach agreement on the shape of their final recommendations or plan. This end result probably would be a hybrid of the original proposals and the coalition and/or compromise proposals developed during the "game" time.

Theme Six: Communication

If you can't communicate, the least you can do is shut up.
—Tom Leher

The concept of communication can be used as an escape for not dealing with real problems. When a disagreement occurs in class concerning foreign policy or students' rights, someone is always quick to say that what we need is more communication. Usually what is needed is more understanding.

I want to speak briefly about communication, not as a problem-solving mechanism, but as a concept that has some objective and recognizable traits. Good communication is a tool, a learned technique that allows its master to better influence his/her destiny and future.

The first basic tenet of communication is that one has not communicated until a response is received. In the theological aspect of this point, God has communicated his love to us in many ways, but it means nothing until we respond to that love. A teacher may be very articulate about a particular period of American history, but if the students do not respond by relating this information to their own use, it has no meaning. Every speech, every conversation, every attempt at communication is incomplete until it becomes two-way.

A second point to remember is that we always use indirect as well as direct communication. I may be telling you verbally that I am concerned that you did not receive the grade you expected on the exam, but my general composure may be communicating that I really don't care. Communication goes beyond what one is saying verbally; it includes the other meanings that are "hidden" in mannerisms, voice inflection, and enthusiasm level.

Finally, verbal expression is only one form of communication. What we say is often overshadowed by what we mean. Non-verbal communication is increasingly more important, as students begin to experience the value of being sensitive to the needs of others and having others be sensitive to their needs. The "Micro-lab" described below concentrates on non-verbal communication. Do not view this as an introduction to "sensitivity training"; it is merely one more way of teaching this valuable aspect of the total concept of communication.

Communication is intangible, yet I can hold it; communication is abstract, but I can be supported by it; and communication is elusive, but I am freed by it.

THE NAME OF THE GAME:

Values/Youth Culture

This simulation game is an unpublished game developed by the author. It is designed to be played in six to eight hours on an overnight retreat for a group of eight to thirty persons.

I am including a discussion on VALUES/YOUTH CULTURE in this chapter for two purposes—the first is to illustrate the relationship between values and communication; the second is to point out that you can use simulations to communicate and teach abstract theory and that the teaching and learning of theory can be made fun and can become the organizing principle for a youth retreat or a planning weekend.

In order to communicate how VALUES/YOUTH CULTURE attempts to illustrate an abstract theory of values, it is necessary for the reader to have some knowledge of the abstract theory being illustrated. VALUES/YOUTH CULTURE is based on a theory that has been developed by Dr. T. Richard Snyder, of the staff of Metropolitan Associates of Philadelphia. It is called "The Transformation of Society" and is a theory about how changes in values occur. With the permission of Dr. Snyder, I am reprinting an unpublished paper called "Some Notes on the Process of Transformation" (c.f. Appendix Two).

PURPOSE:

VALUES/YOUTH CULTURE is designed to show how positive social change can occur using explicit values as the basis for designing and implementing change. The change that an individual or a group wishes to accomplish need not be universal or national in scope; it can often—in fact usually—be restricted to the local or particular level.

The theory that sets forth a model for changing from a war-time to a peace-time economy can be the same theory that provides a model for changing from an adult-controlled to a youth-controlled church-youth program. The emphasis on this game is on creating a process that implements the explicit values that underlie the organization seeking change.

THE PLAYERS:

Although the game can be played by a group that consists totally of young people or adults, a mixture of the two is more ideal. The two roles in the game are youth and adult, and it is suggested that some youths play adults, some youths play youths, some adults play adults, and some adults play the role of youths. The game director should use his discretion and knowledge of the personalities involved in choosing who is to play what role.

ENVIRONMENT

The process steps for VALUES/YOUTH CULTURE follow the outline of "The Process of Transformation," which is printed in Appendix Two. Because this simulation differs from most games that have been packaged and published, I am going to be more descriptive about how it is used with a group.

The game is structured to give the leader maximum flexibility in determining what the actual content of the simulation will be. The first task is to determine what needs to be changed in the environment in which the group functions; for example, I am going to describe the game VALUES/YOUTH CULTURE with the objective of changing the authority for youth-programming in a local church in order to give youths more control over the content and the allocation of resources available for this programming. In your situation, the problem may be quite different. The primary concern could be how to develop leadership in the senior high youth group; how to improve the quality of the programming in the youth division of the church; or how to communicate the values of the youths to the rest of the church. The important thing to remember in using this game is that it provides steps that can be utilized to plan for positive change based on explicit values.

To create a context that would break open the question of power in youth-programming, I constructed a fictional situation that is defined by the following "issue statement":

> The senior high fellowship has just issued the following ultimatum to the Board of Education of the church: "Due to the perverted and inadequate value system of this church, which is reflected in its inept educational program, we demand that all members of the Board of Education resign within the next month. Until a more satisfactory way can be found, the cabinet of the youth fellowship will take over the functions of the Board of Education."

At this point the game players are divided into groups of eight. Each group of eight is then subdivided into groups of four.[1] In each total group of eight, one subgroup is identified as the adult representatives to the Board of Education and the other group is identified as the leaders of the senior high fellowship.

From this point, the groups of eight will function independently of one another and preferably should be given separate rooms. At the completion of the simulation, the different groups will reconvene to share the results of their work. However, for the duration of the game, they will function as autonomous units.

The first task within each group is to become "conscienticized." The first step in "conscienticizing" is to overcome fatalism. The leader should explain that quite often individuals feel that there are given structures, given ways of doing things that are unchangeable. Usually

---

[1] The subgroup's size can vary from three to five.

this is referred to as "tradition" or "that's the way it's always done." At this first step it is the job of the leader to communicate to the groups, either through pep talks or dramatically, that they do have the power to make any change they deem necessary within the present institutions. There is nothing about the present order of power concerning the youth fellowship that has to remain constant. It is possible for the group to turn complete authority over to the youths or to withhold any authority from them. In other words, there are no "givens" with regard to what change is possible or not possible; all the structures and resources are the property of the game players.

The second step in "conscienticizing" is "the recognition of incoherence." The youth and adult subgroups begin to identify the incoherences by reconstructing the present reality. Each subgroup, with materials that the leader provides, is instructed to create a collage which describes how the youths and the Board of Education presently relate to one another in terms of control over youth-programming. The collage should be impressionistic and a group effort. Allow approximately 30 minutes for the collage creations. The two subgroups then return to the total group to report on their collages. As each subgroup explains to the other groups the reality that its collage is designed to reflect, they are instructed to list on separate sheets of newsprint (one sheet for the adults, one sheet for the youths) the values that are implicit in the collages; for example, values such as cooperation and participation will more than likely be implicit in both collages. At this point, use the "ends values" which are listed above in the outline on "The Process of Transformation" as a check. For instance, is "economic justice" or "collective interdependence" among the values that are implicit in the collages? [2]

After the collages have been explained and the values implicit are listed, both are plastered on the wall in each team's working area. The teams now individually examine the present way of ordering and they must list what they wish to keep in a new way of ordering, and what they wish to throw out in the new ordering of power. For instance the youths may decide that they want the adults to have some participation in youth programming; however they do not want the adults to decide how money is spent within youth programming. The adults, on the other hand, may also decide that they like participating in decisions concerning youth programs but do not want to be in the position of having veto

[2] The technique that I suggest for checking these values against the ends values listed in "The Process of Transformation" is to merely list the title of the ends values in that outline on a piece of newsprint and post it permanently in the room where the simulation is taking place. From time to time refer to this list, defining when necessary, to supplement the values that are growing out of the simulation. The check against the ends values in "The Process of Transformation" should be an informal, rather than a formal, task for the game participants.

power over programs and allocation of resources. During this listing process, it is important that the team members really dig into their roles as youths and adults and list a number of things that they feel negatively about with regard to the present power relationships.

The negative aspects of the present reality are used to illustrate the third step in conscienticizing; namely refusing to accept ongoing patterns. The refusing to accept ongoing patterns is the link with the second major component in the outline, "creativity."

Still working in teams, the two subteams are asked to list on individual sheets of paper the new things they would like to see in a new structure. The act of "looking at the new" is identifying some pieces of a new reality. Building upon the positive list that was previously made, the youth team may want to add things such as adults have less power and procedure for giving non-church youths power in decision-making. The adult team may list things such as exposing youths to the responsibility of power and making them aware of other aspects of church life such as worship and building maintenance—areas where they could also petition for changes. There are many ways in addition to the paper and pencil methods that can be used by the teams to illustrate the "new ideas" that are being generated. Materials should be available for them to make posters and to draw illustrations of the new ideas. Paper of different colors with different colored magic markers can be used to decorate the team's "turf" with manifestations of its creativity and new ideas.

The second step in creating is the putting of these ideas into a comprehensive proposal which may serve as a solution.

In the sharing of the new ideas that have been created within the teams, they are asked to add to their list of values. The youths may now have decided that one of their values is a multi-racial youth fellowship. The adults may want to add leadership training for youths and adults to their list of values (again the values that are added will grow from the work the two teams have been doing and can be checked against the list of values from the fourth section of "The Process of Transformation"). After sharing the new ideas and listing additional values, the two teams return to their respective working areas and begin the task of drawing up a comprehensive proposal for the changes that they wish to see implemented.

In "wooing new combinations" the teams should draw upon the positive list that they made in the previous section as well as the new ideas that were generated in the total group sharing session. The proposal or proposals that the two teams are working on need not have steps of implementation. The important thing in this section is the creation of a workable new idea that does not contain the unacceptable patterns of the old way of ordering power relationships.

After the two teams have completed their proposal or proposals (this can be done in thirty to forty-five minutes) they convene together to share them. Through debate, negotiation and general discussion, the group should come up with one new proposal, which must represent a realistic shift in power from the present way of ordering things. For example, on one run of this particular game, the group decided to establish a youth board in the church, which would have control over all youth-programming and allocation of resources for youth-programming. The group may have a little trouble arriving at a workable solution. However, with a little gentle prodding from the leader and a reminder that it does not necessarily have to live with the results of its decision, the group should be able to come up with a proposal.

Once the group has decided upon the proposed change it wishes to see implemented, that change must be institutionalized. Institutionalizing a new idea is the heart of planning. [3] Since the group is now unified on one idea, it is important to reinforce that unity of purpose.

For the task of institutionalizing, re-divide the group in half; however, this time place two youths on each team and two adults on each team. They should be advised to maintain their roles as youths and adults, but to consider themselves joint strategy teams.

One strategy team is assigned the task of working out the theory of the proposed change. It should respond to such questions as "why is it necessary?" and "how will it work?" The second strategy team will have the task of developing the method of implementation. This involves setting forth a step-by-step process for making the new idea a reality. The method of implementation should include whom to contact first, second and third and what groups should be called upon for support and when. Each stage of the proposal of implementation should have a time frame. It is important that the strategy team on implementation not be permitted merely to put down conclusions or end points, but that it sees the process of implementation as a series of steps, each building on the other. In reviewing the work that the group has done on institutionalizing the proposed change, emphasize the meaning of institutionalization, which is that the idea can be projected into the future, irrespective of the individual or groups who generated the idea.

The preceding description of VALUES/YOUTH CULTURE should be complete enough for you to use the game in your own setting. As with most simulation games, there is plenty of room for your own creative variations on the game and also for revision.

GENERAL COMMENTS:

In designing or using a game like VALUES/YOUTH CULTURE, there is a great deal of latitude allowed for the number of participants and the amount of time used for the game. It is my feeling that, for best results, a game such as this should be used on an overnight retreat or a planning weekend in order to give the group some leeway in the amount of time it spends on each of the process steps.

The timing of this particular game is less exact than with the more structured simulations. The leader will need to "sense" the progress

[3] For a more detailed explanation of one particular planning model, Strategic Planning, see the chapter on Happiness.

that the group in making and move the group at its own pace. This does not mean that it should be given as much time as it asks for at each step. The idea is to keep the game moving and all the participants involved almost all of the time.

Something else to remember when using this game is that the variations possible are limitless. For example, the first step of the first section on "conscienticizing" suggests that each team make a collage. You could also ask each team to write a play, a five-minute dramatization, or to create a human sculpture, or to make a watercolor, or you name it.

One variation that I have used successfully comes as an added step at the end of the game. If the group has moved quickly and there is time remaining, it can build upon the theory of the new idea and the strategy for implementation by developing ways of packaging and selling the idea. For instance, I have divided the group into three new teams and asked one team to develop an outline of a position paper on the proposed change, a second team to design a multi-media presentation, and a third team to create a dramatic demonstration of the need for the proposed change. Again I am sure that you will think of other things that can be added at this point.

You may wonder how many different groups, that is, autonomous simulations, one leader can manage. I think it is ideal to have one leader with each group. However, if that is impossible, one leader can manage as many as three simultaneous simulations. The thing that the leader loses in leading more than one simulation is a fairly thorough understanding of the ideas that are being proposed.

NOTES

THE NAME OF THE GAME:

Micro-Lab

Playing Time———————— 1½ to 2 hours

Number of Participants ——— 14-20

Although a micro-lab is not a simulation game, the exercises involved are often referred to as games and have some of the same qualities. I am discussing a micro-lab in this chapter because of its pertinence to the issue of communication. What follows is a description of one micro-lab experience; there are many more and there are many variations on the micro-lab that I describe.

PURPOSE:

The micro-lab is designed to speed up human interaction. It is a group of exercises that, taken together, intensify and make more meaningful human relations among the participants.

THE PLAYERS:

The micro-lab described below has been used successfully with junior highs to senior age citizens; however, junior highs have some difficulty comprehending the concepts of interpersonal relations and some elderly people have problems with the physical part of the exercises.

A micro-lab is designed for people who have a relatively healthy self-concept and for groups that have a positive understanding of their identity and task. In other words, a micro-lab is not a therapy group and is not designed to deal with individual emotional problems.

Some people who participate may have emotional problems. However, there is plenty of opportunity for them to drop in or out of any exercise. The leader does not really have to worry about someone going off the deep end as a result of a micro-lab; however, one should be sensitive to the feelings of those who seem to be holding back or are having a difficult time with the exercises.

DESCRIPTION OF A MICRO-LAB:

The series of exercises are divided into three sections. The first deals with "establishing one's identity in relationship to the total group." The exercises are as follows:

Exercise Number One: The participants wander around the room for three minutes, eyes open but not speaking. They are directed to sense the presence of others by seeing, touching, feeling and smelling. Ask the participants to "sense" who else is there.

Exercise Number Two: With eyes closed, each participant is asked to stand still and reflect upon "how do I perceive others in the group, and how do others perceive me?" Two minutes should be allowed for this silent exercise.

Exercise Number Three: The third exercise is a repeat of the first except this time the participants have their eyes closed. Approximately three minutes should be allowed for this exercise.

Evaluation: The members of the group are asked to pair off (preferably with someone that they do not know well) and discuss how they felt about the first three exercises. No more than five minutes should be allowed for this discussion.

The second set of exercises can be referred to as "establishing small group identity." The participants are divided into two equal groups and instructed to stand in a circle.

Exercise Number Four: Each group is directed to begin playing with an imaginary ball for three to four minutes. The emphasis is on flexibility and creativity. This exercise, like the other exercises in the small group, is done non-verbally.

Exercise Number Five: This exercise lasts for six minutes and is called "trust circle." As a participant is so moved, he steps to the center of the circle, folds his arms, closes his eyes and, with feet together in one spot and knees stiff, goes limp. It now becomes the task of the rest of the group to support this person and give him/her a feeling of trust.

Exercise Number Six: This exercise also should take about five minutes. It asks that an individual member of the group, as he/she feels ready, go to another individual in the group and express, non-verbally, "how I feel about you right now." The pressure time in this and the preceding exercise is used to encourage participants to act quickly and not procrastinate. The leader should make the participants aware of the time factor.

Evaluation: Each small group now takes five minutes to discuss the three previous exercises and is asked to observe the following three basic rules:

1. Talk in the first person. That is, when referring to what has just taken place don't say "you know how it feels"; say "I know how it felt."

2. Talk about the here and now. Do not refer to what happened yesterday or last week that has some bearing on this, but talk about what is happening now.

3. Talk to another person in the group, not about another person. When referring to a specific incident or a specific feeling about another person, talk to that person directly, rather than to another member of the group. This third rule can be abbreviated "don't gossip."

The final set of exercises begins in dyads. The two small groups are broken up and the participants encouraged to pair off with another person—preferably someone that they do not know well. In this final section of the micro-lab, the emphasis is on building a total group.

Exercise Number Seven: Sitting back to back on the floor, the participants are instructed to communicate non-verbally the emotions of fear, love, hate, joy and frustration, using only their backs as a means of communication. Thirty to ninety seconds should be allowed for the expression of each of the emotions mentioned above.

Exercise Number Eight: Facing one another for a period of ninety seconds, the participants are instructed to communicate, using only their eyes. This exercise, like the previous one, is to be done non-verbally.

Exercise Number Nine: After a three minute discussion in the dyads concerning the two previous exercises, each pair is directed to join with another pair, to form groups of four. They continue their discussion for the next three minutes. Then the groups of four are directed to join with another group of four, to form groups of eight. After the discussion has continued for another three minutes, it would probably be necessary to point out that a group of four has been left out of the new pairing.[1] The two groups of eight or the three groups of four (depending on the number of participants in the micro-lab) are directed to decide, within their group, what they would like to do with the group of two or four, and then

[1] The number of participants in the micro-lab will determine how this next exercise functions. If there is an odd number of participants, the leader should participate to make it even. If there are 14 participants, they should move directly from the forming of groups of 4 into deciding how to incorporate the pair that was left out. If there are 20 participants, groups of 8 should be formed before deciding how to incorporate the group of 4 that has been left out. In the event that there are 16 participants, draft 2 to "help you with administering the final set of exercises." Briefly explain that they are to serve as observers and make comments on how the other participants react to the last set of exercises.

do it. The group that has been left out is directed to decide how it will react to actions taken by the two groups of eight. From this point the micro-lab is controlled by the participants and the realignment, the pulling and tugging, the in-group—out-group dynamics, are always significant.

Evaluation: At least thirty minutes should be allowed for a general discussion and an evaluation of the entire micro-lab experience.

NOTES

THE NAME OF THE GAME:

The Road Game

Order from _____ Herder and Herder
232 Madison Avenue
New York, New York 10016

Cost _____ $1.45 for the booklet

Playing Time_____ 45 minutes - 1 hour

Number of Participants _____ 16-32

PURPOSE:

Like *Starpower* and *They Shoot Marbles, Don't They? The Road Game* can have many valid interpretations. It can legitimately be used as an international simulation, a group dynamics exercise, or a governmental systems game. However, one aspect of the game is consistently present: the participants must encounter one another in a profoundly meaningful way if their team is to accomplish its objectives.

The game is best used as an exercise in leadership training and the development of communication skills. The participants divide themselves into teams and choose a leader. From that point on, all communication among the teams must go through the leader.

Although the objective of the game is to complete a specific task, the primary learning comes from an examination of the process used to complete the task.

COMPONENTS:

The participants divide themselves into four teams, each of which has control over an equal amount of territory (represented by equal-sized squares). The teams have to build roads (painted lines on the squares) from their own territory through the territory of one or more of the other teams. Obviously, a certain amount of negotiation must go on in order for any team to build a road.

The game also provides for the assignment of differing characteristics to each of the four groups. For example, one group is asked to be aggressive and competitive, while another group is instructed to seek to accomplish its goals through "enlightened self-interest" with an emphasis on a respect for the rights of others.

Taken together the game components produce a creative mixture of human interaction and realistic problems in communication.

THE PLAYERS:

The game can be played by as many as thirty-two persons, but works best with sixteen to twenty-four. As the number of participants increases, the group maintenance problem increases.

At the beginning of the game the leader reads a fairly long, but not necessarily complicated set of rules that must be followed. These rules are not written down, and they are not repeated once the game has begun. The players must remember the rules, apply these rules as they heard them, and mediate any disagreements that may occur as a result of two differing interpretations of the rules.

The players are very much on their own during the course of the game session. The necessity for the players to govern their own setting is essential.

ENVIRONMENT:

The game should be played in a large open room, where the participants can gather in teams around a playing board that is four feet square. The room should be arranged in such a way that the participants can easily arrange themselves into a comfortable position for discussing the game at the conclusion of the twenty-minute gaming session.

An important part of the simulation is a hearing that takes place after the twenty-minute "road building" exercise. During this hearing the groups make decisions about which roads are valid and which are invalid. Again the emphasis is placed on the process used to make the decision.

In leading the discussion that follows the game, the teacher should focus the students away from the win/lose aspect of the game into a discussion of how this game helps us to understand the ways in which we relate to one another.

The game is easy and simple to use and should have a positive effect on the students' attitudes toward one another. In order to maximize the learning of communication skills, I would suggest that one de-emphasize the role-characteristics (competitiveness, idealism, etc.) of each of the teams.

NOTES

THE NAME OF THE GAME:

The Desert Survival Problem

Order from _____ Human Synergistics
Detroit Trade Center Bldg.
Suite 1022 Executive Plaza
Detroit, Michigan 48226

Cost _____ $2.50 for the Leader's
Guide and 50¢ for each
Participant Folder

Playing Time_____ Approximately 3 hours

Number of Participants ____ 32-48

PURPOSE:

Group Problem-Solving/Decision-Making is the most important aspect of this game. People who participate in this simulation are rewarded for consensus decision-making and group problem-solving. The Instructor's Manual says, "the Desert Survival Problem is really a test of how well a team can gather and utilize the various bits and pieces of information that each has on the subject."

In addition to the focus on team behavior, the game also is significantly concerned with individual awareness. The individual participant performs the same exercise that the full team performs. Therefore, it is possible to compare one's individual effectiveness with the effectiveness of the total team.

Since the issue in this game is survival, the player also becomes aware of his/her ability to make decisions in a crisis. The question of surviving in a desert with few resources is something that (according to the authors of this game) can be measured objectively. After solving the problem to the best of their ability, the individuals and the teams have a better understanding of how to work as individuals and a group to solve a life-and-death problem.

In addition to learning how to maximize group decision-making, the game also teaches the participant some important information about the desert and about survival in a crisis. At the end of the Instructor's Manual the story of Bishop James Pike's death is analyzed. It points out all the things that he and his wife could have done to survive that they did not do.

COMPONENTS:

The primary component is the imminent life-and-death crisis. The participants' ability to think clearly, reserve energy, and react rationally to the crisis is essential to survival.

The other primary component of the game is the resources available for survival. Each team has fifteen items that must be ranked in priority—individually and as a team. The way a team ranks these items will determine how long that team would be able to survive.

THE PLAYERS:

As many as forty persons can play the game comfortably. The materials are designed for eight teams of five people. Although the best team size is five, one can use as few as four or as many as six on a team. However, it becomes quite difficult to monitor the whole process and keep the pace of the game moving with six on a team. Usually it is better to use extra persons as observers rather than increase the size of the teams.

ENVIRONMENT:

The Desert Survival Problem was designed primarily for use in industry. Since so much work today is done in teams it has become necessary for corporations to train their employees to work together. The same type of cooperative effort that improves production in industry can also improve working conditions in a school, a church committee, or a governmental task force.

The environment for this game should project openness and warmth. Both of these qualities are much too abstract to describe, but they can be discussed.

A large room with space between the teams gives the participants a feeling of openness. In such a setting the players tend to be less self-conscious about being overheard by someone at the next table, or on the other teams. Easy access to the coffeepot and a comfortable seating arrangement can communicate an atmosphere of warmth. Because the exercise is likely to expose a participant's rigidity and point up the cold reality of his inadequacies in logic, the environment should say, "you have room to express yourself; you are among friends."

After spending an hour solving the problem as a team, the analysis of the decision-making process begins. The team scores and individual scores are rated against the rating of an expert. The emphasis is not on how well each team scored, but the process used to reach that score.

The game provides some helpful suggestions on group process and communications skills. Although individual results can reflect negatively on a person's ability to think logically or communicate clearly, there is room in the simulation for limited modification of behavior.

The strength of this game is that it brings many of the issues of communication to the fore; it does not, however, deal with or solve our

216

communications problems. I feel that *The Desert Survival Problem* has much to add to the initial stages of a unit on communication and cooperative learning, but do not expect it to do more than introduce the issue. A lot of good data will be generated by this game; the instructor will have to follow up the game by formulating this data into a constructive learning experience.

NOTES

THE NAME OF THE GAME:

Generation Gap

Order from ——————— Western Publishing Co.
School & Library Dept.
850 Third Avenue
N.Y., N.Y. 10022

Cost ——————— $15.00

Playing Time——————— Approximately 1 hour

Number of Participants —— 6

PURPOSE:

"GENERATION GAP is a simulation that involves its players in conflicts between superiors and subordinates. Such conflicts are endemic to life. Teacher, student, boss, worker, officer and enlisted man—each contains this potential for conflict." And, therefore, each contains the necessity for compromise and mutual appreciation. This game assumes the natural conflict between parents and teen-agers and attempts to construct an environment in which this natural conflict can be mediated.

According to the authors of the game:
The game is structured to encourage
—each player to learn that some issues are less important than others and are worth trading off to the other;
—parents could recognize that teen-agers have the power of their own disobedience and therefore cannot be forced to obey simply through threat and heavy punishment;
—teen-agers to recognize that continual disobedience of parental orders is self-destructive in the long run;
—parents and teens to communicate on all the issues working toward compromise.
(instructor's manual, p. 6)

One can see from this statement of the game's purpose that it is designed to be used with parents and teen-agers who are already in communication. In other words, its best use is with parents and teen-agers who have a healthy relationship as opposed to those parents and teens who have an unhealthy or non-communicative relationship.

COMPONENTS:

The game GENERATION GAP is built around five issues. They

range from: should the teen-ager be allowed to wear his hair long to should the teen-ager be home by 10 p.m. from a date? They do not really represent the "gut" issues that divide parents and teen-agers. The weakness of the issues represents one of the weaknesses of the game. This weakness, however, can be easily corrected by merely having the participants write their own issues and place them on the cards.

Once an issue is placed on the table each parent-teen pair discusses it in an attempt to reach an agreement. Since the issues are constructed so that the parent and the teen-ager are on opposite sides, an agreement is possible only when one of the two parties gives in to the other.

If agreement is not reached the parent has the opportunity to order the teen-ager to respond in the manner that the parent chooses. In response to an order the teen-ager must "behave." He/she may choose to follow or to reject the parent's order.

The final component of the parent-teen relationship that is formally treated in this game is "parental supervision." The teen-agers are not required to indicate to the parents what their behavior is on a particular issue. Since the parent and teen-ager are treating five issues simultaneously the parent cannot supervise them all. On those that the parent "supervises" he/she must decide upon punishment when this behavior is discovered.

The structure of the game focuses on the end results: did the child behave or misbehave? There is too little emphasis placed on the processes that lead to the final decision.

THE PLAYERS:

The game is only designed for six players: three parents and three teen-agers. If one were to time/cost this game it would mean that each individual would be paying $2.50 per hour; that is too high for GENERATION GAP.

In the game the parents compete against other parents and the teen-agers compete against other teen-agers. Although the game can be scored in a way to choose the "best family," the primary competition for the participants is with their peers. The game designers state no preference as to whether or not parents and teen-agers of the same family should be paired off together. They do state that participants should be allowed to choose their own partners. It is my opinion that the game works better when parents are paired with the teen-ager from another family. Since the purpose of the game is to learn more about one's self in relationship to a family member of another generation, I feel this can better be accomplished when the atmosphere is not as emotionally charged as it would be between a parent and teen-ager from the same family.

ENVIRONMENT:

The action in GENERATION GAP is more subdued than the traditional simulation game. The six participants spend the majority of time

220

interacting with their trans-generational counterpart. And there is no structured way in which participants can compare notes among teens.

The idea behind GENERATION GAP is sound and the material is very attractively presented. (This game represents the best packaging job that I have seen.) However, a worthy topic and good packaging does not offset the basic weakness of the game, which is that GENERATION GAP does not provide a framework where its participants can find clues to solutions to the problems of the "generation gap." Parent and teen-ager are pitted as adversaries, and points are scored by one giving in to the other or by making a positive response to an authoritarian command.

This competitiveness points to another weakness of the simulation game. Parents and teen-agers are encouraged to compete against one another, and the game operator is instructed in the directions to declare a winner after the game has been completed. Since I believe that one of the primary strengths of educational simulations is the opportunity to move away from win/lose type games, I feel that the competitive aspect of GENERATION GAP undermines what little educational value the simulation has.

However, GENERATION GAP is not a game to be avoided. The teacher who uses the game should encourage the participants to rewrite the issues and, finally, to create a mechanism for more group interaction. I would suggest that the game be stopped after each round and the six participants discuss and evaluate the decisions that each has made during the previous round. Also, I can see no positive value in "keeping score."

My suggestion is that the person who uses the game should declare "the winner" to be the participant who can come up with the best revised form of a simulation game on the "generation gap."

NOTES

THE NAME OF THE GAME:

The Poverty Game

Order from _____ Rev. Jim Egbert
Pilgrim United Church of Christ
4418 Bridgetown Rd.
Cincinnati, Ohio 45211
Printed in: *Colloquy* Vol II, No. 3
March, 1969

Playing Time_____ 1 hour

Number of Participants _____ 15-20

PURPOSE:

Each person possesses a unique background of experiences which shapes his/her values and determines the perspective from which all new experiences are seen. Communication between two people or two classes of people often becomes impossible because two different sets of life experiences have different values and visions. "Where are you coming from?" becomes a question whose answer must be understood before communication can take place, but that understanding often is impossible simply because one person cannot share another's life experiences.

Jim Egbert's THE POVERTY GAME is one of the shortest and simplest ways to help middle class and/or poor people cross the communication gap and begin to grasp why their values and visions are different, not only after an hour-long common experience, but after many years of life in the same country.

The game provides an experience of poverty in the midst of affluence and the dynamics which that situation sets in motion. Role-playing is a part of the game, but the participants do not have to play their roles very hard. In fact, the unequal distribution of wealth which starts the game usually determines how everyone else views them—no matter how they decide to play their own role. Values and visions become the natural results of that set of experiences which become a part of each person while he/she merely tries to understand what everyone else is doing.

COMPONENTS:

The game operator plays the mayor role for the community that is simulated in POVERTY. The police chief, a representative from the welfare department, and a representative of the church are designated at the beginning of the game and usually carry a symbol of their roles, such as a badge for the police chief. They act stereotypically like conservative, God-fearing, law-and-order Christians and uphold the laws of the town, frown on poor people, protect the stores from theft, hassle idlers and vagrants (the poor people in the game), and, in general, keep up the conservative image of the town. A jail is set up for lawbreakers according to the power structure's definition and is invariably occupied during the game. The church representative usually tries to solicit money from the rich to support the church or to give money to the poor (always in another country). Energetic church representatives have had successful hymn-sings in jail, prayer-meetings in the middle of town, and collected lots of money for irrelevant purposes from the rich people. The welfare representative can give allotments to the poor if they are unable to find jobs and can succeed in registering them properly. Police chiefs have been known to extort protection money from wealthy stores and citizens. Mayors have been known to tax irregularly to rid the town of an undesirable element such as the junk store.

About halfway through the game, a player who has been observing is sent into the town as an "organizer" with a loosely defined role of organizing the poor people for their common betterment. He is sometimes called a poverty-program person but is not explicitly told what to do and is not identified to the other players who, by this time, know everyone else in the town. Organizers can act as liaisons with the city officials to get services for the poor people. They can also organize demonstrations and boycotts of the town's businesses.

In addition to the institutional components, the participants represent the different social classes because they are given unequal resources with which to complete the common task. Other players are storekeepers and represent the commercial component of the game.

The all-important task (succeeding) is represented by making a collage. All participants are extended the carrot of recognition if they complete a prize-winning entry.

ENVIRONMENT:

Much activity happens simultaneously: collage-makers buy (or steal); police protect; harassment begins; storekeepers encourage spending and compliment the wealthy on their choice of purchases; some are jailed; welfare workers give out small allowances. At the end of the game the collages are displayed and the participants are allowed to jeer or cheer.

The game materials (the elements of a collage) are readily available; there is no charge for the game instructions, and the advance preparation is negligible. For all of these reasons, it would make a great ice-breaker for someone who has always wanted to be a game operator, but never had the courage or the opportunity.

APPENDICES

Appendix One

HOW TO DESIGN YOUR OWN SIMULATION GAME

The task of designing and using your own simulation game may be easier than you think. The primary ingredients are a clear objective, a concise but general analysis of the issue, a gimmick for creating dynamic action and an atmosphere where learning can take place.

The simulation game, a unique learning tool, has two basic parts. As a simulation it contains a structure, an artificial environment, and a representation of reality. The structure may be as rigid and complex as the moon-landing simulator used by astronauts to prepare for their "real landing on the moon" or as simple as a table placed in the middle of the room representing a school cafeteria. The second irreducible part of a simulation game is role-playing, in which the participants represent an idea, a group, or defined persons. This gaming aspect of the simulation game adds the dynamic to the simulated setting, and they jointly create the context for this unique learning experience.

When I am asked the question "How do you design the simulation game?" I respond, "Let's write a game." When I do this I follow a very simple four point outline: 1. choose an objective; 2. identify the major systems that relate directly to the issue that is stated in the objective; 3. specify one problem to solve that typifies the issue identified in the objective; 4. impose a structure on the game.

Using this simple outline I'm going to describe how I believe the game *Ghetto* was written and also describe how I worked with a group of trainers and teachers to write the game "Flash back, Flash forward." The primary objective of *Ghetto* is to give participants an experience of the frustration and difficulty of living in a modern urban ghetto. The primary systems that relate directly to the issue set forth in the objective, namely living in an urban ghetto, are schools, the employment market, the welfare system, illegal activity and the desire to improve the neighborhood. Each of these systems is represented in categories on the game board which establishes the simulated environment. The particular problem choosen which relates to the issue set forth in the objective is "making it." In *Ghetto*, making it is defined as obtaining enough money to live while improving one's education in order to be able to qualify for a better job, thereby earning more money. Finally, the structure that is imposed on the simulated problem is a role definition for ten participants, a playing board which includes a representation of the major systems, and a mechanism (chance cards, dice, indicators on a neighborhood conditions' chart) for relating the participants to the various systems. When packaged, these ingredients comprise a very useful educational experience.

Describing how *Ghetto* may have been written is, however, still a distance from describing how you can design your own simulation game. To move one more step down the abstraction ladder, let me describe how I designed a game for a group of trainers and teachers at a conference I recently attended. The group for whom the game was being designed had been together for a total of three weeks over the last eighteen months. It had participated in three different week-long training conferences. This particular one was the final conference in the three-part series, and the game that was designed was to be used to close the conference.

We decided that this simulation game would have two objectives. First, there would be a closing experience for the participants and, second, it would be a disbursing experience with emphasis on reentry into their normal day-to-day activities. Since these two objectives form a type of paradox, it was determined that we would construct a game for each objective and that the two games would run simultaneously.

The primary systems that bore on the first objective (the closing experience) were three different small-task groups. The systems that were to be dealt with upon leaving the conference were home, the office, and the world which is represented in the newspaper headlines. The problem we chose to focus on in the closing simulation was "How to recapture the significance of the small-task groups in a meaningful way." For the reentry simulation the problem became "How do I react in my initial contact with the real world?"

It should be pointed out that the objectives, major systems, and the specific problems were all drawn together during a group session. As the trainer I merely played a catalytic role and helped by giving ideas about how a simulation could be constructed from the data that was being presented.

The structure that was imposed to make these two simulation games one total experience was a combination of ideas from other group leaders, a committee of participants, and my own "sense" of what would work. I think by describing the simulation game that took place one can "get a feel" for how the design emerged.

However, before describing what happened let me point out one other aspect in designing a game. Be realistic about the time constraints and the constraints imposed by the physical setup. In the particular simulation I am describing we had an hour and a half block of time and one large room in one building and three smaller rooms in an adjacent building. When planning a simulation be careful to detail the amount of time that is needed for each step. Also see that tables, chairs and other paraphernalia are either removed or arranged in a manner that contributes positively to the simulation environment.

In the simulation game "Flash back, Flash forward" that was designed for the conference closing experience we used the large room for game number one, "Flash back." The three rooms in the adjacent building were used for game number two, "Flash forward" or how do I reenter the real world. The structure that was used was one of alternating between game one and game two with a special device for moving from "Flash back" to "Flash forward."

The simulation game was begun with the participants seated in the small task groups in which they had worked the last three days of the present conference. The task of each of these groups was to prepare and present a training package to a local client group. In the simulated context I, as game operator, indicated that the time was the afternoon prior to their formal presentation and someone in the group has just said "This thing is going to be harder than I thought." For seven minutes the participants relived some of the anxiety that went into the preparation of their respective presentations. Precisely at the end of seven minutes I indicated that they were to gather in groups of three for the car ride home. Physically arranging the triads with two in the front seat, one in the back and having them walk in a semi-seated position, each small group of three "drove" to one of the three rooms in the adjoining building. During their "drive" home they discussed the conference they were leaving behind and what they anticipated finding at home.

Each of the three rooms in game two was set up to represent one of the three systems: home, office, and the world of the newspaper headlines. The participants were divided equally among the three rooms. The first room was set up like a kitchen. One of the participants was instructed to create a role-play between a husband and wife, the husband having just returned from a fantastic experience at a "design skills conference." He wanted to share this experience with his wife but did not know how, and the wife seemed more concerned about what had been going on in her world during his absence. The second group was an office where the conference participant reentered a world (role-play by the other conference participants) of daily office routine. The role-play in the third room was a subway in which the participants who were commuting to work got into a dispute over the current major news items which happened to be the President's trip to the People's Republic of China. In each of these role-plays the conferees have an opportunity to identify with the person who is reentering "the real world" as well as experience vicariously the feelings of those who have not "retreated" into the conference setting. The role-play session lasted for ten minutes and the participants returned to game one, where they were arranged in the task groups in which they had begun the week's conference. Here the setting was day two of the conference and one of the group members had just said, "Putting together a presentation for this local church group looks pretty easy to me." For seven minutes the participants relived their initial feelings of confidence and cockiness. In order to provide some comic relief and a bit of physical exertion the participants were instructed to "fly an airplane from game one to game two." For the transition they arranged themselves in rows of seats, four abreast on the lawn between the two buildings. Complete with sound effects and flapping motion they "flew" to their destination in game two.

The participants returned to the same room they had been in for the first sequence of game two. For ten to fifteen minutes they debriefed the role-play that had gone on in the previous session. Upon returning for the third sequence the participants were arranged in the task groups that they had worked in during the conference they attended six months previously. The discussion for this period centered around the value of the skills they planned to acquire during that and the succeeding conference. The transition from game one to game two during the third sequence was an "alone walk" between the two game settings. There is a point between the conference setting and reentry into the real world when the individual is completely alone with no one and no group to rely on. Whether it be the long, lonely six-hour bus trip home or the ten-second walk from the car to the front steps of the house, there is a period when the individual must "get himself together."

When the participants entered game two for the third sequence they were asked to share their anxieties about reentering the real world and to reinforce the positive aspect of disbursal.

Although that completed the simulation experience the conferees returned to the setting of game one, made a human sculpture and did a snake dance around the building. The objectives of the games were achieved because the participants were emotionally disengaging themselves by "walking back through the experiences that had brought them together" and simultaneously reinforcing one another's feelings of anxiety and adequacy in reentering the real world. They were at once being thrust apart by projecting themselves into the past and into the future—hence, flash back, flash forward.

I have gone into some detail in describing this particular experience to illustrate how easy it can be to design your own simulation game by just following a simple four point outline. To recap I would merely say that you be clear and concise in stating the objectives and analyzing the issues; be rigid about imposing time constraints and constructing the simulated environment; and, finally, be flexible in creating the procedure for the simulation game.

Appendix Two

Some Notes on
THE PROCESS
OF TRANSFORMATION

OUTLINE OF THE PROCESS

I. *Conscienticizing* (The process of transforming consciousness)

    A. The first element is the recognition of the man-made nature of the current ways of ordering and structuring our reality and relationships. This is the break from a fatalistic view of the world in which the systems and structures are assumed to be given and fixed. It is the beginning of man's recognition of his potential to grow and to change, and by extension, that the structures can grow and change. Even more, it is the beginning of one's understanding that he himself is a potential shaper of the future. (P. Friere, C. W. Mills)

    B. Another element in the transformation of consciousness is the recognition of incoherence.

        Incoherence is a contradiction within a system or way of arranging things, an anomaly which doesn't fit the paradigm. It is the recognition that the current way of ordering things is either not in keeping with the statements of intention (e.g. policy, values, constitution) of the ordering group, or that the systems of connection are breaking down; they do not function as they once did because of the introduction of new facts or data.

        Awareness of incoherence can produce community. As persons begin to recognize a commonality of oppression, alienation, etc. they often move toward one another as a means of dealing with their situation. At the same time, communities which have formed for other reasons may serve the function of a problematizing context for some of their members. Incoherence both drives people to each other and it is sometimes discovered in a community where openness is encouraged.

    C. A third element in the transformation of consciousness is the refusal to accept the on-going patterns and structures.

        The intolerableness of the incoherence leads one to recognize that to play according to the old rules is no longer possible.

        The transformation of consciousness is more than awareness. It involves willing. It is when one refuses to adjust, refuses to accept the on-going patterns, that the element of will enters the picture. Until one refuses to accept the existing patterns as the basis for the future, his awareness is too partial for transformation to occur.

II. *The Development of Creativity*
   A. The minimum level of creativity is the ability to ask questions about the new.

      There are two sets of questions which need to be asked, viz. the shape of the new and the implementation of the new. Questions about the shape of the new involve clarification of the values we wish to incorporate, a critique of the available options, and an analysis of the components which need to be considered.

      The second set of questions relate to the policies of implementation: systems analysis, strategy alternatives, motivational concerns, means values, recognition of operational realities, basic human factors, etc.

   B. Maximum creativity goes beyond the "reality" of the above questions and involves the "wooing of new combinations," i.e., devising new conceptual frameworks, new paradigms, etc., which deal with the answers to the above questions.

      "Wooing new combinations" grows out of new beliefs about man and how he relates to his world. It involves a shift in values or, more frequently, in the priority ordering given certain values. (A new combination also involves new ends-means relationships.)

      Wooing new combinations involves the free use of imagination and utopian thinking.

      Wooing new combinations grows out of wishing. But, it goes beyond wishing. It involves the highest degree of intention or self-conscious willing. As R. May says, "on this dimension, will enters the picture not as a denial of wish but as an incorporation of wish on a higher level of consciousness."

      Wooing new combinations can best occur when planning long range. Long-range planning has the option for more freedom from the parameters of the present. Technocratic planning, on the other hand, tends to be short range. It reflects the time-bias of industrialism. (A. Toffler)

III. *The Formalizing of the Power for Transformation* (The process of institutionalizing the new possibilities and combinations)

   A. The need for institutionalization

      Since this is a mass, technological society, it necessitates even more the institutionalizing of relationships due to the systems of interconnections.While a certain degree of tribalism or small community may be possible in certain aspects of life and society, it is impossible to avoid complex interrelationships. Because of the necessity of these interrelationships it is necessary to institutionalize. We need to take our creative ideas and individualized or parochial actions and make them social.

B. The style of institutionalization

Institutionalizing forces us to the critical question of freedom vs. control. Technical systems designs threaten to become the new form of institutionalization. By their very nature, they probably will not institutionalize the power for transformation because they depend so heavily upon predictability. "To the extent that we increase predictability and performance reliability by selecting predictable and reliable components, to that extent we reduce the system's freedom and its capacity to deal with emergent situations effectively." (R. Boguslaw)

C. The means of instutitionalization must be done in a participatory rather than an elitist or centralized way. This is for at least two reasons: First, it is consonant with our concept of justice. It is the end we desire. Second, in an increasingly complex and rapidly changing society, decisions must be decentralized in order that they get dealt with. (A. Toffler)

The greatest problem is that the institutionalization of the new must go on in the context of the old, or in the struggle with the old. That is because most of the changes we make are partial and they are set in the total systemic context.

THE SHAPE OF TRANSFORMED SOCIETY

In developing our understanding of the Process of Transformation, we have attempted to speak of how we might move to a more fully human society and individual man. That section deals with what are sometimes referred to as means values. But it is not sufficient to speak of how we are going to bring about the new: we must also have some ideas about the shape of the new.

While it is necessary and possible to begin to sketch the kind of society we want, it is equally necessary to avoid locking into fixed principles as if they were universally and eternally unchanging and understood. Since the history is an open process, the shape of the future can never be exactly predicted nor can any explanation of what we now desire be the final word. What we predict or state as our goals and ends values are shaped out of a tension between the current situation in which we find ourselves and the vision we have.

That vision may change and be reshaped in ways we cannot anticipate. Nonetheless, it is important that we have some clues or directions in order to guide our movement. The main thing to keep in mind is that the shape of the new society is and must continuously be an open thing. It is as the future continues to break in upon us that we shall shape and reshape, or reinvent its shape.

With this in mind we offer the following as some of the basic guidelines or directions for the shape of the future society.

We believe the new society should be built upon:

1. *The Provision of Basic Material Needs for All People.* Everyone shall have the right to the basic needs of an adequate and nutritious diet, healthy housing, quality health care and life-long education. Wealth shall be distributed on the basis of human needs and the rights mentioned above. Material needs must be dealt with internationally, rather than on a nationalistic basis. All structures which foster economic exploitation by individuals, organizations and nations should be eliminated.

2. *Ecological Wholeness.* The community of which we are a part includes the entire natural world and we must order our lives in a way which takes this seriously. Planning, resource usage, production and consumption shall be based upon the recognition of the interrelatedness of all life.

3. *Eroticization of Life.* Man needs to affirm the totality of his humanity: the affective as well as the rational, grief as well as happiness, play as well as work, the tender as well as the tough, imagination as well as pragmatism.

4. *Maximum Participation of All People.* All persons shall have the right to participate in the decisions which affect their lives and in which they choose to participate. The decision to grant that authority to someone else or to some organization needs to be based upon a coincidence of values, and should be chosen rather than coerced or manipulated. Further a means of recalling the granted authority must be built into the decision-making structure. Self-determination must be held in tension with the achievement of collective values.

5. *Cooperative Interdependence.* The myth of individualism must give way to the recognition of our interdependence with every other person and nation. Genuine individuality, nationalism and small collective identities need to be held in tension with the totally collective nature of life in a complex technological world.

6. *Pluralism.* Man should enjoy and encourage a diversity of values, cultural forms and life-styles for the richness which can result from their interfacing. The myth of the melting pot is a reductionism to a conformity based upon the dominant culture.

7. *Openness to the Future.* All structures and principles need to allow for the possibility of growth and change. We need structures which evolve as open systems: relationships which are entered into in such a way that the power for continuous questioning, critique and reshaping is possible.

Appendix Three

SUPPLEMENTAL LIST OF CURRENTLY AVAILABLE GAMES

FREEDOM

ARGOS

Urban political process

Betty Zisk
Department of Political Science
Boston University
Boston, Massachusetts

BEYOND A REASONABLE DOUBT

Simulates the experiences of a jury member

J. Malcolm Moore
Jack Rabin
Athens, Georgia

CONSENSUS

A game of electoral strategy

Players: 2 - 4
Time: 2 hours +
Cost: $7.95

John Reed Koza
Scientific Games Development Corp.
Box 427
Ann Arbor, Michigan 48107

DEMOCRACY

The workings of the legislative process

Players: 6 - 11
Time: ½ - 4 hours
Cost: $8.00

Western Publishing Company
850 Third Avenue
New York, New York 10022

INTERCITY SIMULATION LAB

Simulation of teaching in the inner city classroom, a teacher training tool.

Time: Several sessions
amounting to
about 10 days

Science Research Associates, Inc.
259 East Erie Street
Chicago, Illinois 60611

PRESIDENTIAL NOMINATING CONVENTION

Politics of presidential nomination

Playing Time: 1st cycle—
2 - 3 hours
additional
cycles—
1 hour or
less

Science Research Associates, Inc.
259 East Erie Street
Chicago, Illinois 60611

SIMSOC	Illustrates the nature of social order and process of social conflict and social control

Players: 20 - 50
Time: 6 - 8 one-hour
      sessions
Cost: Player's manual $4.95
      (forms included)
	The Free Press
	Dept. F
	Riverside, New Jersey 08075

SUNSHINE Players: 20 + Cost: $10.00	Racial problems in a typical American city Interact P.O. Box 262 Lakeside, California
WOODBURY POLITICAL SIMULATION	Election simulation Little, Brown and Company 34 Beacon Street Boston, Massachusetts 02106
NEW TOWN Players: 7 - 10 Time: 3 - 4 hours Cost: Varies by level of complexity	New community development Harwell Associates, Inc. Box 95 Convent Station, New Jersey 07961

LIFE

BALANCE Time: 3 hours Players: 20 + Cost: $10.00	The ecological dilemmas of four families Interact P.O. Box 262 Lakeside, California 92040
DIRTY WATER Players: 2 - 4 Time: 1 - 2 hours Cost: $7.95	Introduction to the problems of water pollution and ecological balances Urban Systems, Inc. 1033 Massachusetts Avenue Cambridge, Massachusetts 02138
EDGE CITY COLLEGE Players: 15 - 30 Time: 3 - 4 hours Cost: $30.00	The problems and alternatives facing higher education today Urbandyne 5659 South Woodlawn Avenue Chicago, Illinois 60637

PEASANT	Agrarian village life; its structure, purpose, and customs
	Peter Hamon c/o Luis Summers University of Oklahoma Norman, Oklahoma
SMOG	Introductory board game about air pollution and city management
Playing time: 1 - 2 hours Number of players: 2 - 4	Urban Systems, Inc. 1033 Massachusetts Avenue Cambridge, Massachusetts 02138
URBAN NEWCOMER GAME	The experience of the rural migrant in a city
Number of players: 20 or more	G. Gutenschwager, Associate Professor of Planning School of Architecture Washington University St. Louis, Missouri
WALRUS I Players: 15 - 30 Time: 4 hours Cost: $75.00	Water and land resources use simulation Urbex Affiliates, Inc. 474 Thurston Road Rochester, New York 14619
F.L.I.P. Players: Up to 30 Time: 2 - 8 hours Cost: $34.00	Family Life Income Patterns Instructional Simulations, Inc. 2147 University Avenue St. Paul, Minnesota 55114

HAPPINESS

BEDROOM BACCARAT	Population planning
	Harvard Center for Population Studies 9 Bow Street Cambridge, Massachusetts
COMPACTS (Community Planning and Action Simulation)	Planning game focusing on problems of social welfare, health, and mental health
Players: 20 - 60 Time: 3 hours - 3 days	Robert Wesher c/o Random House 201 East 50th Street New York, New York 10022

EDPLAN

Players: 29 - 36
Time: 2 hours
Cost: $25.00

Educational planning game

ABT Associates
55 Wheeler Street
Cambridge, Massachusetts

FUTURE

Players: 4 - 12
Time: 1 hour

Predicting and planning for the future

Olaf Helmer and Theodore Gorden
Institute for the Future
Riverview Center
Middletown, Connecticut

LAPPAGE

Latin American planning game

Beth Simon
27 Garland Drive
Amherst, New York

POLICY NEGOTIATIONS

Players: 6 - 10 optimum
           (more possible)
Time: Primary game—
      2 - 3 hours
Cost: $60.00

How to influence the decision-making process

Urbex Affiliates, Inc.
474 Thurston Road
Rochester, New York 14619

P.O.G.E.

Deals with the inherent conflicts in the practice of city planning

Francis Hendricks School of
Architecture
California Polytechnic
San Luis Obispo, California

COMMUNICATION

CULTURE CONTACT

Players: 20 - 30
Time: 2 classroom
      periods
Cost: $25.00

Misunderstanding and potential conflict between two cultures

ABT Associates
55 Wheeler Street
Cambridge, Massachusetts 02138

STRIKE

Players: 28 +
Time: 3 hours or more
Cost: $10.00

Labor-management relations

Interact
P.O. Box 262
Lakeside, California

YOUTH CULTURE GAME
Players: 20 - 80
Time: 2 hours
Cost: $15.00

Introduction to cross-generational communication

Urbandyne
5659 South Woodlawn Avenue
Chicago, Illinois 60637

PEACE

CZECHOSLOVAKIA 1968

Power dynamics in Warsaw Pact, interparty relationship, ideological conflict, foreign policy.

John P. McAbee
Department of Political Science
Boston University
Boston, Massachusetts

DESTINY

Players: 25 +
Time: About 3 hours
Cost: $10.00

Cuban crisis of 1898

Interact
P.O. Box 262
Lakeside, California 92040

DIPLOMACY

Players: 2 - 7
Time: 5 hours
Cost $8.00

Strategies of foreign relations and war

Games Research, Inc.
48 Wareham Street
Boston, Massachusetts 02118

GRAND STRATEGY

Players: 11 - 30
Time: 1 - 3 periods
Cost: $25.00

International relations and diplomacy in pre-World War I Europe

ABT Associates
55 Wheeler Street
Cambridge, Massachusetts

INTER-NATION
SIMULATION

Cost: $53.95

International relations, trade, alliances and war

Science Research Associates, Inc.
259 East Erie Street
Chicago, Illinois 60611

LIBERTE

Players: 20 +
Cost: $10.00

French Revolution

Interact
P.O. Box 262
Lakeside, California

MIKEGASIMO

Milton Keynes economic gaming-simulation model

Robert Sarly
6 Willoughby Road
London N.W. 3, England

MISSION	American foreign policy in Vietnam
Players: 20 +	Interact
Cost: $10.00	P.O. Box 262
	Lakeside, California

SQUATTER	Delphi exercise on squatter communities in Latin America
	Jose Villegas
	College of Human Ecology
	Cornell University
	Ithaca, New York

WORLD GAME	The maximization and optimization of all world systems
Developed by	
Buckminster Fuller	Tom B. Turner
	Spaceship Earth Exploration by Design
	Science
	P.O. Box 909
	Carbondale, Illinois 62901

LOVE

HELPING HAND STRIKES AGAIN	The problems faced in trying to help
	For information write to:
	Dr. Fred Goodman
	School of Education
	University of Michigan
	Ann Arbor, Michigan

RISK	Simple military game
Players: 2 - 4	Parker Brothers
Cost: $4.95	Can be purchased at department stores

TO "B" OR NOT TO "B"	The controversy over grades
	J. Malcolm Moore
	Jack Rabin
	Athens, Georgia

Appendix Four

THE NAME OF THE GAME:

COG

Order from: ———————— J. J. Mar-Tam and Associates
1053 Delamont Avenue
Schenectady, N.Y. 12307

Cost: ———————— $20.00 (with script)
$25.00 (with tape)

Playing Time: ———————— 2 hours

This game is concerned with
the Reorganization of the Federal Process
and with Revenue Sharing.

PARTICIPANT'S GUIDE

INTRODUCTION

Cog means—Coalesce, Oppose or Grapple.

Upon reading this brief introduction you will be a member in good standing of the Middle County Citizens' League. The League is an assortment of "good government" people from around the county who are currently concerned about the future of government's participation in the delivery of human services. This simulation will take you through a crucially historic period; namely, the advent of revenue sharing.

The format for the simulation is three meetings of the Middle County Citizens' League (MCCL) on October 30, 1972, January 30, 1973 and March 30, 1973. During each meeting, the MCCL will divide into five Task Forces: Community Action, Education, Law Enforcement, Urban Renewal and Transportation. You will be assigned to one of these Task Forces and will remain in that Task Force throughout the game. Some participants will be tapped to serve on the County Board of Representatives.

Each MCCL meeting constitutes a round of the game. The fourth and final round will consist of a meeting of the County Board of Representatives on April 14, 1973.

Following is a brief description of the county in which you will reside for the next two hours:

Middle County is situated eighty miles from the Capital. The total population is 242,000 people. Although most people live in Jefferson City, the population is distributed through rural and suburban communities.

The economy of the county is supported primarily by two major companies: P. O. Lutant Chemicals, which employs 9,000 people and Farmers' Delight, Inc., a local plant that makes manure spreaders and snowmobiles.

The black population is just under 10,000 persons and most of the black population is located in Jefferson City. Although minor riots broke out following the death of Martin Luther King in 1968, the city has been relatively quiet. According to federal figures, 18% of the population falls below the poverty line.

Each round you will receive specific instructions concerning the unfolding action of the simulation. With the exception of some of you who will be assigned particular roles by the Chairman of MCCL (the Game Director), you are free to role-play creatively within the framework of the game.

There are only two simple rules you need to follow:
    (1) sit loose with the action; and
    (2) have fun.

GAME STRUCTURE

MCCL County Board

Round I, October 30, 1972

Orientation	A.	Orientation
----------------------		----------------------
Consideration of Task Force Reports	B.	County Manager makes presentation on General Revenue Sharing

---

Round II, January 30, 1973

Special Meeting on the President's Proposed Budget	A.	attend MCCL meeting
----------------------		----------------------
Task Forces meet to consider effects of proposed budget on their areas	B.	Board meeting to allocate General Revenue Sharing Funds

---

Round III. March 30, 1973

General meeting to review Revenue Sharing	A.	attend MCCL meeting
----------------------		----------------------
Task Forces meet to prepare presentation for or against the resolutions coming before the County Board at their April meeting.	B.	Board members will circulate among the Task Forces, quietly campaigning for reelection.

Round IV. April 18, 1973

The County Board of Representatives hold their regular monthly meeting with the MCCL members present and each MCCL Task Force makes a brief presentation on the three resolutions that are being acted on by the Board.

242

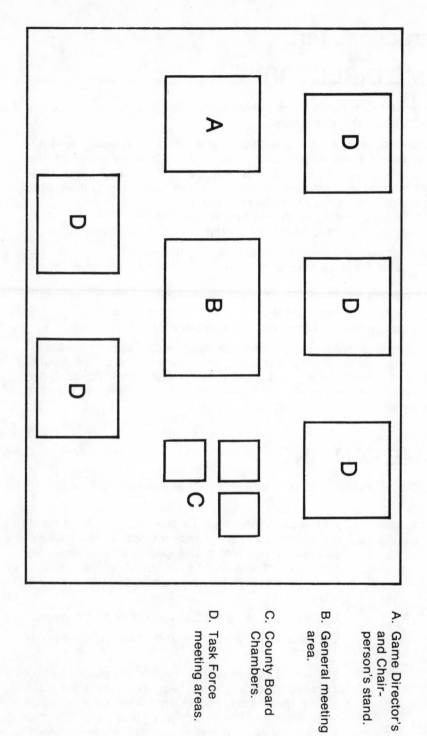

ROOM SET-UP

A. Game Director's and Chairperson's stand.

B. General meeting area.

C. County Board Chambers.

D. Task Force meeting areas.

Instructions for:

GAME DIRECTOR

If you are an experienced simulation-gamer, you will be amazed at the ease with which this game runs.

If you have never played a simulation game, and furthermore are frightened to death at the prospect of running one, don't be alarmed! The process is really quite painless. The participants do the work; you merely point them in various directions at specified times during the game. There are only two simple rules to remember:
1. always be concise in your instructions; and
2. do not forget to have fun yourself.

Before beginning the simulation session, you should:
1. take about an hour to familiarize yourself with the game material
2. set up the room and pass out the material for the first round (Participant's Guide and the First "Citizen's Voice")
3. set up the tape recorder and listen to the full tape.

You should also make the following role assignments prior to the beginning of the game: County Manager, 3 to 5 members of the County Board of Representatives, and 2 persons for each Task Force whose roles are explained in the Round I material. Name tags are contained in the game kit.

GAME OUTLINE

ROUND 1. Introduction to the game and the five issue areas

Time: 15 min. A. All the participants will be seated in the main meeting room facing the speaker's stand. The moment you stand up before the group you become the Chairman of the Middle County Citizens' League. Others will recognize your role by the name tag you wear.

As Chairman of the MCCL, you will begin the game:

1. Welcome everyone to the monthly meeting of MCCL.

2. Remind everyone that the primary purpose of tonight's meeting is to hear the reports from the five issue Task Forces that were established in September. The issue Task Forces are: Community Action, Education, Urban Renewal, Mass Transportation, and Law Enforcement.

3. Tell the group that before they break up to hear Task Force reports they may be interested in a news feature that WGBR carried this evening on our organization. Turn on the tape. The first beep indicates the end of the Round I portion of the tape.

4. Dismiss the participants to their respective Task Forces and the County Board members to the Chambers of the Board of Representatives.

Time: 15 min. B. Instruct each Task Force to choose a chair person and ask that person to read the report of his or her respective Task Force.

Instruct the County Manager to make the report on General Revenue Sharing to the Board of Representatives.

Allow each of the groups about 10 minutes to discuss the reports.

If time permits, each group should write a two-sentence Statement of Purpose.

Some questions that you may be asked:

What are other details concerning our particular project? They are free to *create* additional details that do not conflict with the information they already have.

Why are we only considering one specific project in each Task Force? To examine an entire area would be impossible within the scope of this game. Our purpose is to understand revenue sharing and that is best accomplished by isolating how it affects individual projects.

Do we need to know what the other Task Forces are doing? Not necessarily. Their work should become apparent as the game progresses.

ROUND II. JANUARY 30, 1973, The President's Proposed Budget

Time: 15 min. A. You are again chairing the monthly meeting of the MCCL. The announcement of the President's Fiscal Year 1974 budget has aroused the passions of a number of people. Therefore, you have decided to devote the entire monthly meeting to a consideration of the proposed federal budget.

245

For the general session, play the tape of WGRB's weekly news conference in which two of our local representatives react to the proposed budget. (TURN ON THE TAPE. The beep will signal the end of this section.)

Instruct the participants to return to their Task Forces.

Time: 15 min. B. In the individual Task Forces the respective Chairmen will read a report from the MCCL's research staff on how the proposed budget will affect the particular program that Task Force is studying.

The County Board of Representatives will first hear a report from the County Manager on how he proposed that they spend their general revenue sharing money. After a brief discussion they adopt the Manager's plan.

(There may be some who want to make changes in the plan offered by the County Manager, but explain to them that they are simulating Middle County and that in 95% of the counties in America, county governing bodies merely appropriated their general revenue sharing money without much study and even less community or citizen input.)

Time: 10 min. C. Ask each Task Force Chairperson to make a one-minute report on the work of his Task Force.

ROUND III. MARCH 30, 1973, What shall we do about Federal Revenue Sharing?

Time: 15 min. A. The monthly meeting of MCCL will be devoted to a consideration of General and Special Revenue Sharing. As the Chairman you can suggest that they may be interested in excerpts from a tape that was made at a recent revenue sharing conference held in Washington, D.C. (PLAY THE TAPE)

Time: 20 min. B. Each Task Force meets to consider the text of three separate resolutions that will be coming up at the April meeting of the County Board of Representatives. The Law Enforcement group will be instructed to put their major emphasis on the adoption of Resolution III. They should not seek to work coalitions with other groups because they have direct access to the County Board through the majority leader.

The Community Action Task Force will put its major attention on the passage of Resolution 1. Because there is no provision for community action under special revenue sharing, the people concerned feel that their best chance for survival is through the continuation of OEO in its present form.

The other three Task Forces should consider the resolutions on their merit and prepare a three-minute presentation for the County Board meeting.

The Task Forces should be encouraged to try to convince other Task Forces to follow their point of view. Negotiating among the groups can be significant at this point in the game.

The County Board members will be circulating among the groups as observers, but also as candidates for reelection.

As the game director you may have to hurry people a little. Some will be getting tired, others will want to meticulously prepare their presentation. Stick to the time limit and promptly call everyone to begin Round IV.

ROUND IV.   APRIL 12, 1973, The monthly meeting of the County Board of Representatives

All participants should turn their chairs around to face the County Board Chambers. (See attached diagram.)

The County Board Chairman will call the meeting to order and ask the clerk to call the roll. A member of the Law Enforcement Task Force serves as County Clerk. After the roll call, one of the Board members leads the entire group in the "Pledge of Allegiance."

Then the Board Chairman asks if anyone from the public would like to speak. A spokesman for each Task Force steps forth to make a presentation on behalf of his or her respective group. The Task Force spokesmen address themselves to the three resolutions before the County Board.

After the five persons have spoken, the County Board takes up the three resolutions in order and after a brief discussion, votes on each.

SUGGESTED DISCUSSION QUESTIONS
COG   COG

After the game is complete, a discussion of a half hour should be conducted. People should sit in a circle or semicircle together and speak free and informally. End the discussion period while the group is still "warm." Participants should leave the discussion with their minds still turning over events of the simulation.

The questions below are only a guide to stimulate discussion. It is often helpful for the game director to jot down key points and questions as they arise during the game. Also sharing your own observations may serve as a good stimulant to the discussion.

1. Who made the decision on how to spend General Revenue Sharing money? How did citizens participate?

2. Does everyone understand how General Revenue Sharing funds are calculated? Who benefits more, big cities or suburbs?

3. Was the tax reduction legal? (yes) Did it violate the spirit of the legislation?

4. What was the effect of the President's budget on Middle County?

5. Does Special Revenue Sharing bring local people back into the decision making process?

6. What impact will Special Revenue Sharing have on local officials if it is passed?

7. Do you think local officials will be receptive to the new demands?

8. Are coalitions possible on the local level? Who are the necessary participants in a strong coalition in your county?

9. Would it be better to stop Special Revenue Sharing from being passed?

10. How would you respond if one of the groups in your coalition made a separate appeal, like the Law Enforcement Task Force in the game?

COUNTY BOARD

Introduction

During the first part of each of the first three rounds, you will sit as observers of the Middle County Citizens' League. When the League breaks into Task Forces, you will go to the County Board of Representatives' Chambers and hold one of your monthly meetings. During the fourth and final round you will conduct a meeting of the Board of Representatives. Following is an outline of the meetings:

Round I - October 30, 1972
> The County Manager will make a report on General Revenue Sharing.

Round II - January 30, 1973
> You will vote to appropriate your General Revenue Sharing money.

Round III - March 30, 1973
> Since you will all be running for reelection and face primary challenges this spring, you will spend the time during this round casually visiting the Task Force meetings and quietly campaigning for reelection.

Round IV - April 30, 1973
> Three resolutions will be presented for your determination.

GENERAL REVENUE SHARING

The General Revenue Sharing bill recently passed by Congress and signed into law by the President could have a dramatic and positive effect on Middle County. I recommend that the County Board of Representatives review the following details carefully and be prepared to allocate money as soon as possible.

Congress will allocate $5 billion to be spent by the states, counties, and municipalities throughout the country. Each county government receives a lump sum as calculated on a formula based on population, tax effort, and per capita income. As it looks now Middle County will be receiving $1.2 million dollars under General Revenue Sharing for 1973.

General Revenue Sharing is intended to assist localities finance the programs that the local governments deem necessary. Money can be spent for:

—public safety,
—environmental protection,
—public transportation,
—health,
—recreation,
—libraries,
—social services for the poor and aging,
—financial administration, and
—capital improvements.

Money cannot be used on facilities or programs that discriminate against individuals because of race, sex, or national origin, or as matching for other federal programs, or on programs that pay less than the minimum wage. The money must be spent within two years after it arrives.

Also money cannot be spent on education or direct tax reductions.

To summarize, General Revenue Sharing will be arriving soon and I look forward to our county being able to use this money to solve some of our financial problems and to prevent a tax increase.

Respectfully submitted,

Roger Nohow,
County Manager

GENERAL REVENUE SHARING ALLOCATIONS

On Tuesday of this week our office received the first installment of the $1.8 million allotted to Middle County under the General Revenue Sharing Act. This money could not have come at a better time. Not only will the General Revenue Sharing money enable us to meet important obligations, but it will also allow us to reduce taxes by $2.00 per thousand. By applying a portion of the General Revenue Sharing money to highway fund, the $1.6 million will be partially freed for tax reductions.

Below is my recommendation for use of the General Revenue Sharing funds:

Highway repairs	$600,000.00
Garage for Park Department	200,000.00
5% raise for County Employees	600,000.00
Hiring two Sheriff's Assistants	20,000.00
Addition to the County Jail	380,000.00
	$1,800,000.00

Three community groups have sent other recommendations. One suggested financing police officer education at the University ($240,000.00). Another called for creation of a public employment program to help get people off relief. And finally, the Welfare Rights Organization "demanded" that the funds be applied to increase welfare benefits. While these three proposals have varying degrees of merit, I believe my recommendations fulfill the high priority needs of the County.

Respectfully submitted,

Roger Nohow

Roger Nohow,
County Manager

REVENUE SHARING

I.A
SCRIPT

(Tape begins with "Action News" type music)

(Voice over:)

Good evening and welcome to another broadcast of *Challenge*, a special feature of the WGBR News Department which brings you analysis of the news as the news is being made.

(Music continues)

(Voice over:)

This week, *Challenge* is brought to you by the people who produce Soapy-Kids. Soapy-Kids is the only bubble bath that's strong enough to clean your farm pets and your own children, too—all in the same tub of water! Next time you're looking for a bubble bath that's brisk, bracing, baby-bottom soft, and ecologically sound to boot, try Soapy-Kids.

(New voice:)

And: it's fully approved by the U.S. Food and Drug Administration. Soapy-Kids—a division of Kleen Kids, Limited.

(Music up; then fades and concludes)

ANNOUNCER:

This is *Challenge.* My name is Mike Malloy. Tonight we'll be bringing you the first in a series of *Challenge* features devoted to the Middle County Citizens' League. Our series will focus on the current plans and long-term goals of the Citizens' League, the organization whose accent on good government and citizen participation in urban affairs has made headlines nearly every week for the past year.

This week's feature will focus on the history and make-up of the Middle County Citizens' League. We'll look into its constitution, its acting Task Forces, and its plans for the upcoming year.

With me in our studio tonight is Mrs. Edna Modicum, retired President of the Citizens' League. She'll be answering some of my questions about the Citizens' League's efforts. It's good to have you with us, Mrs. Modicum. . . .

MRS. MODICUM:

Thank you very much, Mike.

ANNOUNCER:

In reading the Constitution of the Citizens' League, Mrs. Modicum, I landed upon this sentence: "The League is principally devoted to act as a watchdog for the affairs of local government, bringing the average citizen into close contact with the priority-setting and decision-making processes of all the municipal governments in Middle County." Tell me, Mrs. Modicum, does this sum up the purpose of the Citizens' League?

MRS. MODICUM:

Yes it does, Mike. It's the elected official's job to bring government closer to the people, so we've taken it upon ourselves to bring people closer to the operations of government. The only thing I'd like to add to that quote from our Constitution is the matter of our Task Forces—these are the true "work-horses" of the MCCL.

ANNOUNCER:

Can you tell me how many Task Forces there are and what their topics are?

MRS. MODICUM:

Yes I can. Currently, the Citizens' League is running five Task Forces. Of course that number changes from month to month, since the critical issues facing our local government change quite frequently. But at this time there are five.

The topics of the five Task Forces are Law Enforcement, special programs in Education for underprivileged students, Urban Renewal, a Mass Transportation study for the whole county, and Community Action.

ANNOUNCER:

Tell me what you mean by "community action," Mrs. Modicum.

MRS. MODICUM:

Well, by community action we mean our county's official Community Action Program: Mid-CAP. Mid-CAP is a local office of the Office of Economic Opportunity in Washington, D.C. Mid-CAP is here to assist the poor people in this area by helping them understand and work with the causes of poverty. Mid-CAP has done a *tremendous* job of bringing people through the complexities of the Welfare system, getting them jobs, providing legal assistance, and generally organizing poor people in meaningful ways around the issues which affect their lives.

ANNOUNCER:

Mrs. Modicum, what's it like to be a member of an MCCL Task Force?

MRS. MODICUM:

Well, Mike, I've been on quite a few myself and each one has had a character of its own. I suppose the main thing a Task Force member has to be ready for is hard research. Our Task Force leaders are experts on the topics, so they're always ready to hand out tricky, but worthwhile studies to every Task Force member. Our Task Force on Housing Code Enforcement in 1971 had its members out on the streets 3 nights a week, interviewing tenants in Jefferson City's housing projects. This year's study on Education for the disadvantaged has sent the Task Force to Washington three times to learn about federal grants for education.

ANNOUNCER:

That's very interesting. Our time is running short, I'm sorry to report. Can you tell me briefly what the Citizens' League plans are for 1973?

MRS. MODICUM:

Yes, Mike—I'll try to be brief. Basically, we'll continue the five Task Force studies we're doing now on Law Enforcement, Education, Urban Renewal, Mass Transportation, and Community Action. In addition, the League hopes to begin a full study on the topic of Federal Revenue Sharing.

Revenue Sharing marks a profound change in the way federal funds are passed down to the local level—the money will come to *local government*, instead of coming directly into the hands of local federal agencies. No one is very sure what this will mean for all the federally funded programs in Middle County, but the leaders of MCCL are doing their best to find out as soon as possible!

("Action News" music up)

ANNOUNCER:

And this brings us to the end of *Challenge* for another week. I'm very grateful that you could be with us tonight, Mrs. Modicum. The best of luck to you and the Middle County Citizens' League.

MRS. MODICUM:

Thank you, Mike.

(Music continues)

ANNOUNCER:

Tune in next week at this time, when Soapy-Kids will bring you another installment of *Challenge*, for analysis of the news as the news is being made. This is Mike Malloy, saying thank you, and good night.

(Music up, and out)

BEEP

A monthly publication of the Middle County Citizens' League

Volume III   Number 7   October 30, 1972

TASK FORCE REPORTS READY

After more than two months of diligent work our Task Forces on Community Action, Education, Law Enforcement, Transportation, and Urban Renewal are ready for review at this month's meeting. Each Task Force decided to first isolate what it considered one of the most important projects in the county in its respective area. The Task Force then prepared a brief description of that project.

A brief summary of the Task Force reports that the groups will consider the following:

Community Action: Mid-CAP, the local anti-poverty agency appears to be working well after a long period of fruitless political confrontation and personnel in-fighting.

Education: The More Effective Schools (MES) program in two Jefferson City schools appeared to be working quite well until a group of local parents filed a complaint against the school system charging that the school district had misspent a large portion of the MES money.

Law Enforcement: FIT (Families in Transition) was funded by an LEAA (Law Enforcement Assistance Administration) grant to serve as an indepth counseling service to kids on probation and their families. Although some still consider it a program of "coddling criminals" it has gained widespread political and public acceptance.

Transportation: A mini-bus system serving a large group of heretofore unserved persons (senior citizens, students, and rural residents) is threatened because the OEO grant which subsidizes their operating costs will run out soon and because the bus line has not become self-sustaining.

Urban Renewal: A relatively small urban renewal project in Cowgone has received considerable attention lately because the Democratic County Committee charged that the Republicans had been using the Urban Renewal project for partisan benefit. This particular project is also much in the news because of a plan to build a high-rise apartment complex for low income senior citizens in the center of the project area.

ROLE CHARACTERS
FOR EDUCATION

Chairman: Nancy van der Guard

—Jefferson City School Board member

—sympathetic to all schools in Jefferson City—rich and poor

—adamant about receiving *all* Title I and MES money

—Representative of MES parents group

—a black attorney who favors no funds for Title I and sees unfair "padding" for rich schools

—sends one of his children to a private school

Charles Jackson

ROLE CHARACTERS
FOR URBAN RENEWAL

Chairman: Freddy Cuchinella

—Republican City Councilman

—wants to play down corruption charge and proceed with implementation of the program

—is a strong advocate of low income senior
citizens' housing

—senior citizens' spokesman

—concerned about corruption in
senior citizens' project

—is an old-line Democrat

Clifton Early

ROLE CHARACTERS
FOR COMMUNITY ACTION

Chairman: Richard Everpresent

—Executive Director of the Neighborhood Youth Corp
—1970 MSW
—has gained a reputation for efficiency
—fits the stereotype of the white liberal

—President, Black People's Unity
 Group (BPUG)

—believes that Community Action has
 been corrupted by white liberals

—has been quite active in opposing the
 Mid-CAP family planning project

—maintains a strong commitment to non-violence

Lucille Meriweather

ROLE CHARACTERS
FOR TRANSPORTATION

Chairman: Ms. Laura League

—Vice President of the League of Women Voters

—member of the LWV's group on transportation

—carefully disguised candidate for the County Board

—her husband is very skeptical of her community activities

—has driven a bus for 23 years

—resents the fact that mini-bus
  drivers are paid almost as much as he

—Union representative for MCTA bus drivers

—concerned about the problems of senior citizens

Bus Driver for Middle County Transit Authority:
Howie Handiwork

ROLE CHARACTERS
FOR LAW ENFORCEMENT

Chairman: Chuck Nohow

—older son of the County Manager

—adamant about need for new funding

—young white liberal, whose inexperience shows

—feels no funding should come
for any program to "excuse punk
kids and colored kids from the
jail sentences they deserve"

—has served as sheriff for 32 years

—believes that PAL (Police Athletic League)
is the answer to all problems

Sheriff John Hardcap

COMMUNITY ACTION

The local anti-poverty program (Mid-CAP) was established in 1966 under the Economic Opportunity Act of 1965. Since its inception, Mid-CAP has been hampered by a number of problems. Early confrontation tactics antagonized local government and city officials have not forgotten the marches, the sit-ins, and street demonstrations. The agency also has been the scene of a number of unfortunate personnel battles.

However, despite these problems, the agency appears to be quite vital in the community and it runs a number of programs. Among the many programs are:

Neighborhood Service Centers in seven areas,

Community Organization Staff,

Head Start,

Neighborhood Youth Corps,

Neighborhood Health Center,

Family Planning, and

Training and Technical Assistance.

A small controversy that has received a lot of publicity lately has been the demand by the Black People's Unity Group (BPUG) that the family planning money be refused because family planning is a form of genocide.

Apart from the flare-up over family planning, Mid-CAP is preparing its refunding proposal because its new grant year begins in January, 1973.

EDUCATION

Jefferson City Board of Education has been running programs under Title I of the Elementary and Secondary Education Act for a number of years. But last year they applied for and received a grant to operate two of their schools under a More Effective Schools (MES) program. The program was funded under Title III.

Although some people argued against the concept of compensatory education and others thought that it was too expensive an experiment, the MES program had the full support of the school system. In fact, it was hailed by many as "the answer to our education problems in the inner city."

Indeed, a number of positive changes did take place during the first year of the project. However, a major controversy erupted two months ago when a community group, with the help of legal aid attorneys filed a complaint with the Office of Education. The group charged that services under the MEW grant were incomplete and that some MES funds had been illegally allocated to some affluent schools in Jefferson City.

If the complaint is upheld the school system will have to return to the federal government a sum of money equal to the misappropriated funds. The Board of Education still backs the concept of MES, but the complaint must be resolved before the program can be refunded.

URBAN RENEWAL

Cowgone, a small community of 30,000 has had an Urban Renewal project for three years. The project consists of rebuilding a two block section of the city's downtown.

The project proceeded without much public attention until last February when the County Democratic Committee charged that the Urban Renewal project had become a haven for Republican party hacks and that construction contracts had benefited two major Republican contributors. Although the Democratic Chairman did not charge that there had been anything illegal regarding the project, he did intimate that the employees were using job time for party business and that the costs of the construction contracts in question were exceedingly high.

The town, which is controlled by the Republicans, has tried to ignore the charges and has submitted a routine application for refunding for the fourth year.

The one new item in the fourth year plan is a proposed low income housing development for the elderly. This will be the first project of its kind outside Jefferson City.

TRANSPORTATION

The County Board of Representatives began a new mini-bus line last January that runs through the urban and rural sections of the county. The buses were purchased through a Department of Transportation grant, while the operating subsidy comes from OEO.

The bus line was designed to begin paying its own way by January, 1973, but that is not going to happen. A number of new people are being served, but some type of subsidy will always be needed according to a recent county study.

The value of the new bus line has been the extending of service to three groups of people who have not had access to public transportation before. They are a group of high school students from the rural community of Salvage, who for the first time do not have to walk the 1.8 miles to and from school.

The line also serves, at a special rate, senior citizens in various parts of the county. The seniors like the mini-buses because they are easier to board.

Probably the least noticed recipients of the new service are persons on public assistance who can now catch a bus into the employment office or the Department of Social Services. Previously, most people had to ride a taxi into town merely to keep their required appointments with the employment and welfare departments.

The County Board has been considering dropping the bus service because of its cost, but a small vocal group from the League of Women Voters, who have just completed a study on transportation systems in the County are pressing to keep the service alive.

In order to keep the bus running the County should apply for another Department of Transportation grant to build shelters at the numerous bus stops. They would also need to commit themselves to subsidizing the bus line indefinitely.

LAW ENFORCEMENT

Under a grant from the Law Enforcement Assistance Administration (LEAA) the County probation department has begun a program of family counseling with youthful offenders. The title of the program is Families in Transition (FIT).

FIT has been in operation for eighteen months and after a very stormy beginning now shows real potential. The early difficulty centered around a political battle between the parole division and the sheriff's department. The County Sheriff, with support from some members of the County Board felt that the program was allowing too many "young radicals" to avoid jail sentences. Of the first ten cases, five of the families in the program actually followed through with the counseling.

After about six months, however, the program developed better procedures for determining who should participate, and the political furor died down. Now FIT is quietly moving along with its work, serving about 27 families.

The major issue facing FIT is whether or not the program will be refunded for a third year. Their program year ends in May, 1973, and they may not be eligible for a third year of funding from LEAA.

REVENUE SHARING

II.A
SCRIPT

(Tape begins with "Action News" type music again)

(Then ditto star-section of I.A SCRIPT)

ANNOUNCER:

This is *Challenge*. My name is Mike Malloy. Tonight, in place of our regularly scheduled feature on the Middle County Citizens' League, we bring you an unusual special: a report on the President's Budget Message for Fiscal Year 1974.

When the Budget was announced yesterday, there was an uproar in Washington among liberal Congressmen who expressed shock that the Budget included funding cutbacks for domestic assistance programs.

I'm here in our Washington News Bureau studio right now, and on tonight's *Challenge* we'll be hearing excerpts from two interviews I taped this morning on Capitol Hill. I was fortunate to be able to corner the Representative from the Middle County area, Congressman Al Wright, as well as one of our distinguished Senators, Ms. Senator Judy Leftkowitz.

Before I play those tapes, I'd like to report on the mood here in Washington. The three key topics are Defense Spending, budget cutbacks for Social Programs, and Federal Revenue Sharing.

There has long been a debate here over the extent of American financial contributions to the war in Afghanistan. But the debate easily quadrupled today when the President announced the appropriation of $10 billion for reparations after the negotiated conclusion of the war this July.

Liberals in Congress are further incensed because they believe that program cutbacks in the human services area are coming as a direct result of the increased defense budget.

This is what Senator Leftkowitz had to say on the subject when I spoke to her this morning:

(Crowd noises up)

SENATOR LEFTKOWITZ:

It's ridiculous, Mike, and you know it as well as I do.

Do you suppose the White House knows anything about what this will mean for our state? The budget cutbacks include higher education, anti-poverty programs, rural electrification, Title I of the Education Act. . . .

And not only has he cut back important programs—he has taken the funds which were appropriated by *us* for use in Housing and Child Development, and he's directly *impounded* them. So we not only have a President who is *planning* a budget which underrates human needs—he is *distorting* the intentions and results of a legally sanctioned process: 2 bills passed by Congress.

And where will it all end?

The President says, "Wait until you see the results of Special Revenue Sharing before you criticize the real intentions of this Administration."

Well, I'll tell you what I say to that. I say "Nuts!" to that. We've seen what happened to general revenue sharing in this country: nothing, and in a hurry, too. I see nothing in the proposed legislation for Special Revenue Sharing which makes it any different from the NO STRINGS ATTACHED revenue sharing our municipalities got this year.

The only solutions to the grave social issues facing America and our state are *federal* solutions. Yes, local government should be included in the planning stages for coping with social issues, and yes, local government is qualified to deal with these issues on a hometown basis.

But there is no question that the federal government must be involved in coming to grips with serious domestic problems too.

I believe this Administration is consciously trying to throw out the baby (responsibility for domestic affairs) along with the bath water (shared federal revenues).

ANNOUNCER: (over crowd noises)
Thank you, Senator Leftkowitz.

(Voice is back in the studio)

This afternoon, I was able to speak with Congressman Al Wright, who had this to say about the proposed changes in the federal government. . . .

AL WRIGHT:
(Voice over some voices in background, including street noises)

I've just got one thing to say about all this, Mike: It's long overdue.

Congress has been playing Grandma to the lazy, jobless, welfare-cheaters in this country for far too many years. And I'm in full agreement with the President when he says it's time to stop throwing money at the problems, and pass the money on to our responsible local governments, instead.

I'll admit I was a little surprised to see some heavy cutbacks for our veterans. And after all, these are the men who *did* work to make this country what it is today. Now I think with the war in Afghanistan coming to a close and all those boys coming home, we'll be able to negotiate with the President on this one.

But revenue sharing is the correct route to travel for this country. Any other way means throwing out the baby (that is, the money) along with the bath water (that is, all these grant-in-aid procedures).

ANNOUNCER:
(Voice still outside)

Thank you Congressman Wright.

("Action News" music comes up)

ANNOUNCER:
This brings us to the end of another installment of *Challenge*, the program where we make analyses of the news as the news is being made. Tune in for our next broadcast—a feature study on Federal Revenue Sharing.

(Music up)

(Voice over:)

This is Mike Malloy, saying thank you and good night.

(Music up and out)

BEEP

The Citizens' VOICE

A monthly publication of the Middle County Citizens' League

Volume IV   Number 1   January 30, 1973

NIXON BUDGET PROPOSED DRASTIC CUTS IN HUMAN SERVICES

The proposed Fiscal Year 1974 budget, which will begin July 1, 1973, delivered yesterday to Congress is going to be a bitter pill for many individuals and groups who are directly involved in the human services field.

The budget is only a proposal and is certain to be changed in a number of places by the Congress, but it is a very significant document because it clearly outlines the broad strokes of the New Federalism. More than 100 grant-in-aid programs would be eliminated under his proposal and only part of them would be continued under Special Revenue Sharing.

Following is an overview of how the President's proposals would affect our current Task Force areas:

Community Action: OEO will be abolished by executive order and some of the special programs will be transferred to other agencies. But Community Action, the heart of OEO, will be totally eliminated.

Education: The Landmark Elementary and Secondary Education Act of 1968 would be dismembered. Title I funds, which are for compensatory education for disadvantaged children would be transferred to Special Revenue Sharing. Title III, the source of funding for the MES program would be eliminated. The President has proposed $2.5 billion for Special Revenue Sharing in Education.

Law Enforcement: Unlike many other areas, Law Enforcement money appears to be up. Although the proposed budget would reduce the number of programs, the grant-in-aid programs will remain intact. FIT, however, is in the second year of a special program and it is not clear whether or not federal money will be available for a third year.

Transportation: Money presently available for transportation cannot be used to subsidize the operating costs of local transportation systems. Since the county bus line is presently subsidized by an OEO grant, the expected loss of that money could spell trouble for the program. The money needed for capital improvements seems to be secure in FY '74 from the Urban Mass Transportation program.

Urban Renewal: The Executive budget for FY '74 says that HUD (Department of Housing and Urban Development) would approve no new Urban Renewal programs. Existing programs are eligible for up to 12 months funding. This means the Cowgone Urban Renewal project could be refunded. However, there is no chance that the low income senior citizens' housing project will be built so long as the housing freeze continues.

EDUCATION

The President's proposed budget, if enacted, would probably spell the death knell for the MES program. The President has recommended that Congress pass a Special Revenue Sharing bill on education. Categorized programs funded under Title I and III are not in the new budget proposal.

Money for handicapped vocational training, *disadvantaged*, and impact aid are lumped into Special Revenue Sharing but special programs such as MES funded under Title III would not fare well under the proposed revenue sharing plan.

All is not lost because many feel that Congress will not pass a Special Revenue Sharing bill on education this term and that the Elementary and Secondary Education Act of 1968, which is due to expire June 30, 1973, will be continued for an additional year. In that event, programs will probably be refunded at the same level for another twelve months. This would merely buy more time for the Congress and the President to negotiate a compromise.

Effect of Proposed Budget on:

COMMUNITY ACTION

The 1974 budget announced that OEO is to be terminated. Community Action funds will be cut off altogether by the end of this fiscal year. And existing programs will be phased out over the next twelve months. Head Start would continue to be funded by HEW but it is unclear whether or not the money will come through the CAP agency. Neighborhood Youth Corps will probably be a part of Manpower Special Revenue Sharing, and as such, go to the county. Family Planning and the neighborhood Health Center are eligible for refunding from HEW but it is not clear who will administer the program.

The President of the Board of Directors of Mid-CAP suggests, "If this cruel and callous plan is carried out, our agency will cease to exist. We must resist with all our force."

The MCCL staff recommendation is that the Task Force work to maintain the status quo. That is, oppose the abolition of OEO and oppose the shifting of other grant-in-aid poverty programs to Special Revenue Sharing.

URBAN RENEWAL

Urban Renewal, Model Cities, and almost all other urban programs under HUD would be eliminated under the President's proposed budget. In their place, the administration has proposed an urban community development revenue sharing plan that would give the local governments a block grant.

The key words are decategorization and decentralization. The many urban programs would be decategorized and the power would be decentralized to the local level.

The urban renewal project could continue under such a plan if the town council of Cowgone gave the project high priority and chose to fund it from their block grant.

The Democrats are saying that the Republican controlled Council is sure to continue the project because under revenue sharing, they would have even more power to use it to pass on political jobs and to award overpriced contracts to party contributors.

However, even if the urban renewal project continues, there is no chance for the senior citizens' housing to be built anytime soon. The eighteen-month moratorium on federally subsidized housing has held up this project indefinitely.

TRANSPORTATION

There are two primary sources of money for transportation: (1) the Federal Highway Trust Fund; and (2) the Urban Mass Transit System money. Bus stop shelters for the county's mini-bus line would be funded by the latter grant.

A significant limitation of the transportation money is that it can only be used for *capital expenditures*. DOT (Department of Transportation) money cannot be used to operate or subsidize a mass transit system.

Although there is more than a 50-50 chance that the Middle County Board would receive its grant, it is well to keep in mind that only about one-third of the money originally earmarked for urban mass transit has been spent. When Congress passed the bill in 1970 projecting an expenditure of $3.5 billion, they actually appropriated less money. And a part of the money they appropriated has been impounded by the President.

It is expected that some form of transportation revenue sharing bill will be introduced. There is also growing pressure for more discretion with the Highway Trust Fund and for more flexibility in allowing states and localities to spend money to operate programs.

LAW ENFORCEMENT

Although the Omnibus Crime Control and Safe Streets Act of 1968 is scheduled to expire June 30, 1973, the program money under the Act appears to be intact for the coming year.

LEAA, which runs all the programs under the Safe Streets Act, is scheduled to continue with fewer categorized programs and a little more money. The two grant-in-aid programs—allocation to states by population and discretionary programs—would not be changed significantly under the proposed new budget.

FIT, which is funded from the discretionary grant-in-aid program is now completing Part II of its program. Since this was considered a one-shot grant, there is little chance of refunding for a third year.

The local matching money required for the grant has been 40%, so a significant county investment already exists.

FIT has two primary options:

(1) they could ask the county to pick up the whole program; or

(2) they could alter the program and reapply to LEAA for another demonstration grant.

Although we recommend that they go for the first alternative, there is no need to ignore the second alternative. Both grant-in-aid programs of LEAA will continue and the county could write a whole new program to supplement the work already going on. Also the President has proposed that in FY '74 grants require no local match of dollars. One hundred percent federal funding would remove all financial questions for local officials who want to cut programs like FIT because of their cost to the county.

III.A
SCRIPT
("Action News" type music up)

(Voice over:)

(News introduction, as in other two tapes)

(Music continues; then fades)

Good evening. This is *Challenge* and my name is Mike Malloy. If you have been listening to this program over the last two months, you're aware that we've been concentrating on a subject that's come to be known as the "New Federalism."

We've observed how the proposed national budget for Fiscal Year 1974 will have an unprecedented effect on the operation of domestic assistance programs which affect Middle County. We've seen how many of these programs are now in a state of limbo, as the federal government has let go of its responsibility for running them before the programs have been picked up by local, or any other means of, support.

The key factor for the survival of one-time federal programs may lie in an aspect of this "New Federalism" which has not received much discussion in Middle County to date. I refer to the subject of Federal Revenue Sharing. As you probably know, revenue sharing is the process by which the federal government passes money from its coffers of income tax funds *back* to local governments for their own discretionary use. Many observers of government feel that this process of revenue sharing may be the salvation of local assistance programs *if* local government officials become willing to spend revenue sharing money in this way.

This evening, the *Challenge* program is fortunate to have a recorded broadcast of a speech by Dr. Washington Taxman, an economist who has specialized in the study of revenue sharing and its applications by the Nixon Administration. We now bring you some excerpts of Dr. Taxman's speech. . . .

DR. TAXMAN:
(His voice is heard in a large auditorium: a conference setting with clinking water glasses, occasional coughs, etc. etc.)

The first term of the Nixon Administration gave substantial indications that a move was afoot to *decrease* the role of the federal government in many areas of domestic policy-making. The first formal step toward this "New Federalism"—the implementation of a lasting structure of limited federal involvement in domestic matters—came in the form of a memorandum to Congress in 1970. In this document, President Nixon outlined his plans: (1) to consolidate the Executive Office of the President into four major areas for "Community Development," "Human Resources," "Natural Resources," and "Economic Affairs"; (2) to *simplify* the current grant-in-aid programs for domestic assistance, placing them all under the wings of these four "mega-departments"; and (3) to strive to emphasize local governments as the focus for planning and operating federal assistance programs through decentralization and regionalization, to be accomplished through general and special revenue sharing.

Hence, Federal Revenue Sharing, both general and special, is part of a much larger scheme by the Administration to reduce the direct influence of the federal government in domestic matters, and to pass along both the funds and the authority for governing domestic life to municipal government.

Now just what is Federal Revenue Sharing?

The General Revenue Sharing bill, passed in October of 1972, provides that the sum total of funds allocated by the bill (about $5 billion for 1973) be broken down to states, counties, city and village governments, and town governments, respectively.

The funds are divided among the states and municipalities in accordance with a *formula:*

POPULATION X TAX EFFORT X Average Per Capita Income

Essentially, this formula is used to *compare* municipal governments in terms of the three factors involved, and to divide the funds involved on a proportional basis according to the products which come from the three-factor formula when applied to each municipality. Of course, the factors may change from year to year—population may change, taxes may change, and so forth—so the amount received by a local government may vary from year to year, as well. (Many local governments, for example, have asked whether they would be wise to use their revenue sharing funds to *reduce* local taxes. Now this may sound advantageous from a politician's point of view, but elementary economics will show the politician that if he reduces taxes in one year, his share of the revenue sharing money will *decrease* the following year, since the TAX EFFORT factor will be smaller.)

What happens when the formula-determined check for general revenue sharing arrives in the mailbox at City Hall? The law states that local governments may spend their shares in, and only in, the following two areas: the first area includes the subsections of public safety, environmental protection, public transportation, health, recreation, libraries, social services for the poor and aged, and financial administration. "Area" is called "ordinary and necessary *capital* expenditures which are authorized by law."

There are traditional restrictions on the funds which apply to all current federal grants-in-aid, but these are administrative restrictions which really have no bearing on local priority-setting for use of the funds.

277

In the last analysis, general revenue sharing is a no-strings-attached stipend to the general fund of every state and local government in America. What these governments do with their funds is basically up to the elected officials and the citizens who elect them.

Let me say a few brief words about Special Revenue Sharing, and then we'll close this portion of the conference—I know we're running overtime, so I'll try to be as concise as possible.

Special revenue sharing is still on the governmental drawing board. President Nixon has proposed six broad bills, covering the areas of manpower, urban community development, rural community development, education, transportation, and law enforcement. Under the concept of special revenue sharing, *each* of these six bills would provide a grant to *all* local governments in America. Local governments will be required to spend their shares in the *special area* which each bill defines.

Congressional debate over the six bills has focused on three areas: (a) the formulas for distribution of the money (for example, how does one determine a transportation grant for San Francisco as opposed to one for Minneapolis under the same formula?); (b) the designations or "earmarkings" within each bill (for example, should the urban community development bill include funding for the construction of housing units? Or should the funds be spent on planning, only?); (c) the grand question of whether there should be special revenue sharing bills at all. Many Congressmen have been apprised of the difficulties found in general revenue sharing this year, and some feel that the new special revenue sharing should be withheld until the rough edges of general revenue sharing are smoothed out. Some Congressmen have also stated they would oppose special revenue sharing because they believe the money entailed for municipalities is far less than the money now being received under the grant-in-aid structure.

I hope this introduction to the concept of revenue sharing has helped some of you to better understand the basics of this novel Administration plan. My final message to you is a warning: *Be prepared.* This is an era of rapid change in government as we have known it. And only those leaders who keep abreast of the new dynamics of government as planned by this Administration will have the opportunity to benefit their communities as the proposed changes go into effect.

Thank you very much.

(Applause)

MIKE MALLOY (back in studio)

We've just heard excerpts from a talk delivered at the Mayor's Conference on Federal Revenue Sharing by Dr. Washington Taxman. I hope you were enlightened by Dr. Taxman's presentation. As an amateur on this topic of revenue sharing, I can only echo Dr. Taxman's words of warning: that preparedness and full understanding of new developments in the federal government will be essential if we are to thrive under the innovative laws which are churning out of Congress every week.

This brings us to the end of another *Challenge* feature.

("Action News" music comes up)

I hope you'll be here next week when *Challenge*, the program which analyzes the news as the news is made, brings you a special 4-hour presentation by Dr. Joyce Greenthumb on the biochemical structure of the dandelion seed.

This is Mike Malloy, saying thanks for listening, and good night.

(Music fades; and concludes)

BEEP

The Citizens' VOICE

A monthly publication of the Middle County Citizens' League

Volume IV   Number 3   March 30, 1973

SOME "NUTS AND BOLTS" ON FEDERAL REVENUE SHARING

The flow of revenue sharing funds
(as accomplished by the Department of Treasury in January, 1973, with general revenue sharing funds)

1. From Federal to State governments.

   —shares determined by multiplying POPULATION X TAX EFFORT X p.c. INCOME for each state, and providing R.S. funds in proportion to this product for each state.

2. From State to State government and all local governments within the given state.

   —1/3 of state funds to state government; 2/3 of state funds to local governments.

3. Local government funds are distributed to all COUNTIES within state.

   —shares determined by multiplying POPULATION X TAX EFFORT X p.c. INCOME for each county, and comparing these products to yield percentages of R.S. funds, as in #1.

4. County funds distributed among (a) city and village governments, (b) town governments, and (c) the county government.

   —each of these categories (a, b, c) receives an amount of the COUNTY share proportional to the percentage of all local taxes[1] which it, as a category, collects. (i.e. total city-village taxes vs. total town taxes vs. total county government taxes)

   [1] Throughout this distribution scheme, "local taxes" (or TAX EFFORT) refer to all taxes collected by the locality in question, *with the exception* of those specifically designated as school taxes.

5. From city-village share of the funds, money is divided among cities and villages.

>—funds determined by the POPULATION X TAX EFFORT X p.c. INCOME formula, as accomplished in #1 and #3.

6. From town share, money is divided among all the towns.

>—funds are determined by POPULATIONS X TAX EFFORT X p.c. INCOME formula.

PRIORITY EXPENDITURES

How may a local government [1] spend its federally shared funds?

Under the general revenue sharing bill, the following areas are designated as the "priority areas" in which local R.S. funds may be spent:

1. Ordinary and necessary maintenance and operating expenses for:

    (a) public safety
    (b) environmental protection
    (c) public transportation
    (d) health
    (e) recreation
    (f) libraries
    (g) social services for the poor or aged
    (h) financial administration; and

2. Ordinary and necessary capital expenditures authorized by law.

There are some traditional federal-grant restrictions on the use of general revenue sharing funds. They are:

    (a) the money must be spent within two years after the time it arrives;
    (b) R.S. money may not be used as matching funds for obtaining subsequent federal grants;
    (c) the money may not be spent on projects, facilities, or people that deny the rights of others on grounds of sex, race, creed or national origin; and
    (d) the funds may not be used on projects where people are employed on a pay scale which is lower than the federally prescribed minimum wage.

The only other requirements for governments which receive general revenue sharing funds are that they be subject to a Department of Treasury audit, and that they publish their plans for using the money every six months in widely circulated local newspapers.

[1] State governments do not fall under the restrictions for "priority expenditures" of general revenue sharing funds.

A Planning Council Study on Special Revenue Sharing

Whereas a number of local programs will lose federal funding in 1974 and;

Whereas, the County will probably receive less money under special Revenue Sharing than is being lost in categorical grants; and

Whereas, difficult decisions on allocation of Special Revenue Sharing funds will have to be made by the County;

Therefore, now be it RESOLVED:

that the County Board shall request that the Community Planning Council conduct an in-depth study of Middle County.

The study shall:

(1) Determine needs of residents of the County;
(2) Evaluate all programs affected by the federal budget cuts; and
(3) Make recommendations to the County Board on the use of Special Revenue Sharing funds.

The study shall be completed within three months after the passage of Special Revenue Sharing.

Advantages

—A competent, professional staff will rationally and systematically study the needs and programs of the County.

—The high emotion of the present will give way to a calmer mood when the facts are studied.

—Ineffective programs can be eliminated, while good programs may be retained.

—The community, through the Planning Council, will have an opportunity to participate in the evaluation and funding process.

Disadvantages

—This plan accepts the Special Revenue Sharing Plan even though it may bring less money.

—Funds for some programs will end before Special Revenue Sharing is enacted into law.

—There is no guarantee that the County Board will not ignore the study once it is done.

—The study may cost the County a fair amount of money.

Special Funding for FIT

Whereas, the Families In Transition (hereinafter referred to FIT) has provided a beneficial and needed service in Middle County; and

Whereas, the program does not qualify for a third year of funding from LEAA; and

Whereas, the County Board of Representatives wishes to see this program continue;

Now be it therefore RESOLVED:

that the County Board shall allocate a sum of money for the FIT program equal to the amount that they received last year from the federal government; and

Be it further RESOLVED:

that the County shall continue to fund FIT for an additional twelve months for an amount equal to the 40% matching funds we have been appropriating for the program.

Advantages

—The County can set a positive example by continuing a successful program.

—Family Counseling is a new experimental technique for combating delinquency and it needs more time for testing.

—The Family Counseling program is actually less expensive than maintaining a group of youths in the County Jail.

—If the County does not pick up the funding, the program will end in June.

Disadvantages

—The resolution ignores all other worthwhile programs.

—No systematic planning and evaluation will take place before the money is appropriated.

—If the program is refunded the County Board will show favoritism.

—FIT has not tried to obtain funding from the United Fund and the League of Women Voters.

—Many people still object to this approach of fighting crime.

Opposition to Special Revenue Sharing

Whereas, Middle County will suffer from the loss of funds under the New Federalism's switch from categorical grant-in-aid programs to Special Revenue Sharing, and

Whereas, Middle County cannot afford to lose any federal funding;

Therefore, now be it RESOLVED that:

the County Board of Representatives shall officially notify their Congressman and Senators who represent Middle County and the State that the Board opposes passage of any Special Revenue Sharing Bills.

Advantages

—The categorical programs (Urban Renewal, CAP, Title I, etc.) that are slated to be lost will exceed any projected amount that would be returned under Special Revenue Sharing.

—The increase of local responsibility for programs will increase pressure from favor seekers.

—The strict federal guidelines preventing discrimination, etc. would not necessarily follow if local government ran the programs.

—Physical Development would probably take precedence over social programs.

—The federal government would be passing the buck to local government to cure the social ills, without the bucks to pay for the job.

Disadvantages

—Special Revenue Sharing brings accountability back to the residents of the County.

—Many federal categorical programs have been wasteful and ineffective.

—Some programs, like CAPs, have actually used tax dollars to oppose local government on a number of issues. The County could eliminate the troublemakers.

—Even if fewer dollars are coming, the money will be better spent.

—Many programs had foolish, inefficient guidelines, like requiring a quota of minority employees on jobs.

—People who want change or new programs can get them more easily through the County than by applying to some distant bureaucracy.

A Full Meeting of the County Board of Representatives

April 14, 1973

AGENDA

1. Chairman of the County Board of Representatives calls the meeting to order.

2. All stand for the Pledge of Allegiance.

3. The invocation. Delivered by Pastor Christian Gentile.

4. The County Clerk reads the monthly agenda.

5. Announcement of the three Resolutions to be discussed and voted upon at this evening's meeting.

   I   Opposition to Special Revenue Sharing
   II   Authorization for a Planning Council study on Special Revenue Sharing
   III Special funding for F I T.

6. The five Task Forces of the MCCL make brief presentations, for or against the three Resolutions.

   (1) Law Enforcement
   (2) Community Action
   (3) Mass Transportation
   (4) Urban Renewal
   (5) Education

7. The County Board votes on each of the three Resolutions.

8. Other Business.

9. ADJOURN.